TI-Nspire™

FOR

DUMMIES®

2ND EDITION

TI-Nspire™ FOR DUMMIES®

2ND EDITION

by Jeff McCalla and Steve Ouellette

John Wiley & Sons, Inc.

TI-Nspire™ For Dummies®, 2nd Edition

Published by
John Wiley & Sons, Inc.
111 River Street
Hoboken, NJ 07030-5774
www.wiley.com

WILEY

About the Authors

Jeff McCalla is currently teaching mathematics at St. Mary's Episcopal School in Memphis, Tennessee, where he also coaches the golf team. Jeff holds a bachelor's degree in Christian education with a minor in mathematics from Wheaton College and a Master of Arts in Teaching degree from the University of Memphis. Jeff is the cofounder of the TI-Nspire SuperUser group, dedicated to providing advanced training from the world's foremost experts. In addition, he enjoys traveling the country training teachers as a T^3 Regional Instructor for Texas Instruments. Jeff has had the privilege of being a part of numerous TI-related projects including writing TI-Nspire documents that align with the 2011 edition Pearson textbooks. A highlight for Jeff was receiving the Presidential Award for Excellence in Science & Mathematics Teaching and with it the opportunity to meet President Obama and Secretary of Education, Arne Duncan. When he is not meeting important dignitaries, Jeff enjoys going to ballgames with his boys, playing golf and racquetball, reading Malcolm Gladwell and John Wooden, getting free stuff, teaching Sunday school, and making his wife smile.

Steve Ouellette, or Mr. O, is currently the Math Department chair at Westwood High School in Westwood, Massachusetts. Steve holds a bachelor's degree in electrical engineering from Worcester Polytechnic Institute and a Master of Arts in teaching degree from Boston University. Steve began his teaching career in 1993 after having worked as an electrical engineer at Otis Elevator Company for five years. His engineering expertise helped fuel Steve's passion for incorporating educational technology into his teaching. In addition to this book, Steve has also authored the *CliffsNotes Guide to TI-Navigator* and the *CliffsNotes Guide to TI-Nspire,* and he relishes the irony that he used to avoid writing assignments during his high school and college years. Steve has enjoyed working on a number of other TI-related projects, most notably as a regular activity writer for the *We All Use Math Everyday* program, a Texas Instruments and NCTM joint venture that provides classroom activities that relate mathematics to the TV series *NUMB3RS.* When he's not lobbying for a cameo role on *NUMB3RS,* Steve enjoys spending time with his family, camping, running with his weekend warrior buddies, and watching the local sports teams compete for championships. His passion for baseball is evident in the name chosen for his 90-pound labradoodle, Wally, named after the mascot for the Boston Red Sox.

Dedication

Jeff McCalla: This book is dedicated to my family: my wife, Shannon, and my three boys Matt, Josh, and Caleb.

Steve Ouellette: This book is dedicated to my family: my wife Christine and my three boys, Noah, Ben, and Danny.

Authors' Acknowledgments

Jeff McCalla: I could not have written this book without the help and support of the people at John Wiley & Sons. I want to thank my project editor, Christopher Morris, whose commitment to excellence was instrumental. I also want to thank my acquisitions editor, Amy Fandrei, for showing me the ropes and keeping me to a tight schedule. In addition, I want to thank Douglas Shaw whose sharp eyes and math expertise contributed to the accuracy of the book.

I certainly want to thank my friends at Texas Instruments for their ongoing support. The leadership of Gayle Mujica, Maggie Sherrod, Charlyne Young, and Ellen Johnston has helped make this project possible. In particular, Lydia Neher has come through for me every time I needed an updated device or software build. Of course, Tonya Hancock's support role in forming the TI-Nspire SuperUser group helped fuel my TI-Nspire addiction.

Fellow T^3 instructors have assisted me with their help when called upon. Jennifer Wilson's eye for detail continues to amaze me; her assistance was invaluable. I want to thank Jill Gough, who introduced me to TI-Nspire and changed my approach to teaching. Bryson Perry, Nelson Sousa, and Sean Bird's technical expertise has been a valuable resource.

On the home front, I wish to thank my teaching colleagues, Orion Miller, Sandra Halfacre, and Chrystal Hogan, who were a nice resource when I had questions. In addition to my immediate family, I want to thank my parents, Bud and Elaine McCalla, whose passion for mathematics rubbed off on me. Finally, I want to thank my students, who are the inspiration for much of what I do.

Steve Ouellette: I'd like to thank the good people at John Wiley & Sons who have provided so much support throughout the writing process. First and foremost, I thank my developmental editor, Christopher Morris, for all his support and expertise. I'm absolutely blown away by his knowledge of the *For Dummies* way. I also want to thank my acquisition editors, Amy Fandrei and Tiffany Ma, for helping me adhere to a tight schedule and for all their help related to the content and outline of this book. Additionally, I thank Douglas Shaw, associate professor of mathematics at the University of Northern Iowa, for verifying the mathematical and technical accuracy of this book. There are so many other people that come into play after the acknowledgments are written, so for all the support I know I'll be getting, many, many thanks. Although she hasn't worked directly with me on this project, I must thank

Zöe Wykes for all her help with the *CliffsNotes* books. There's no doubt that my work with Zöe helped make the transition to *TI-Nspire For Dummies* a smooth one.

I certainly wish to thank my friends at Texas Instruments for their ongoing support. In particular, I thank the past and present "big three" of the Content Team — Maggie Sherrod, Steven Bailey, and Vince O'Connell. Maggie, your support and unique sense of humor are so much appreciated. Steven, while your role has since changed, I thank you for being my first "boss" and for your ongoing support. And Vince, you have continued to lend a hand on so many fronts. Every time I need something, whether it's an updated device, a piece of software, or a set of Learning Objects, you've been there to help out. There are many other folks whom I've come in contact with over the years as a result of my affiliation with Texas Instruments. Although I appreciate the professional growth these relationships have offered, it's the personal friendships that I've formed that have made this experience so wonderful.

On the home front, I wish to thank my teaching colleagues and friends for all your interest in and support for this endeavor. In addition to my immediate family, I want to personally thank my mom and dad, Vivianne and Henry Ouellette; my brother Paul Ouellette; his wife, Kate; and their children, Alyssa and Christopher, for simply lending an ear. This means the world to me.

Publisher's Acknowledgments

We're proud of this book; please send us your comments at http://dummies.custhelp.com. For other comments, please contact our Customer Care Department within the U.S. at 877-762-2974, outside the U.S. at 317-572-3993, or fax 317-572-4002.

Some of the people who helped bring this book to market include the following:

Acquisitions and Editorial

Sr. Project Editor: Christopher Morris

Acquisitions Editor: Amy Fandrei

Copy Editors: John Edwards, Jennifer Riggs

Technical Editor: Dr. Douglas Shaw, University of Northern Iowa

Editorial Manager: Kevin Kirschner

Editorial Assistant: Amanda Graham

Sr. Editorial Assistant: Cherie Case

Cartoons: Rich Tennant (www.the5thwave.com)

Composition Services

Project Coordinator: Patrick Redmond

Layout and Graphics: Timothy C. Detrick, Andrea Hornberger, Heather Pope, Corrie Socolovitch

Proofreader: Laura Bowman

Indexer: Sharon Shock

Publishing and Editorial for Technology Dummies

    Richard Swadley, Vice President and Executive Group Publisher

    Andy Cummings, Vice President and Publisher

    Mary Bednarek, Executive Acquisitions Director

    Mary C. Corder, Editorial Director

Publishing for Consumer Dummies

    Kathleen Nebenhaus, Vice President and Executive Publisher

Composition Services

    Debbie Stailey, Director of Composition Services

Contents at a Glance

Table of Contents

Introduction

• •

*T*I-Nspire changes the way that teachers teach and students learn. This transformative device has just undergone a transformation of its own. The TI-Nspire CX features color on the handheld (which is definitely a game-changer). The Touchpad control of TI-Nspire is another innovation that helps students to interact with the mathematics. In addition, the operating system has undergone some major improvements in the last few years. These are some of the reasons we have updated this book.

Many of the improvements to TI-Nspire (both the OS and hardware) and TI-Nspire Computer Software are a direct result of feedback received from teachers and students. Texas Instruments is committed to providing the best tools for the teaching and learning of mathematics and science.

Do you know how to use TI-Nspire to do each of the following?

- ✔ Create and edit documents that contain multiple pages and problems
- ✔ Evaluate expressions in the Calculator application and work with fractional or decimal results
- ✔ Graph and manipulate a parabola
- ✔ Manipulate a geometric object and analyze its changing attributes on a coordinate plane in real time
- ✔ Generate a sequence in the Lists & Spreadsheet application
- ✔ Enter data in the Lists & Spreadsheet application and construct a summary plot in Data & Statistics that compares the data in a side-by-side bar chart
- ✔ Use TI-Nspire Computer Software to insert a color background image on a Graphs page
- ✔ Take pictures of your TI-Nspire Handheld screen and insert them in a word processing document
- ✔ Link TI-Nspire applications to represent information algebraically, numerically, graphically, and verbally

If not, then this book is for you. As you read through the pages, you will find straightforward and practical information that is sure to take you well beyond the beginning stages of using TI-Nspire.

About This Book

This book will not tell you everything you need to know about TI-Nspire. However, I do cover all the basics and give you the tools to start creating your own TI-Nspire documents. Additionally, you will see enough examples to gain an appreciation for the *potential* that TI-Nspire has to offer. It's my belief that your experience gained from reading this book (and playing along on your TI-Nspire device) will give you the confidence to forge out on your own.

I outline a lot of concrete steps and processes to perform a variety of tasks. I use specific math applications as the backdrop for these tasks for the purpose of demonstrating how TI-Nspire can be used as a wonderful teaching and learning tool. As you read this book, you will begin to appreciate that TI-Nspire is a very robust device — if you can think it, TI-Nspire can most likely represent it.

TI-Nspire Terminology

TI-Nspire learning technology comes with its own unique language. The meaning of most TI-Nspire-related words found in this book can be initially understood from their context. However, just to avoid any unnecessary confusion, here are three key terms that you should know right from the get-go:

- **Handheld:** I use this term when referring to the TI-Nspire product that you quite literally hold in your hand. You find three families of TI-Nspire Handhelds: TI-Nspire with Clickpad (the original), TI-Nspire with Touchpad (the next generation), and TI-Nspire CX (new color version), as well as CAS versions of each type (with the computer algebra system built-in). Notice that it is not called TI-Nspire Calculator, but TI-Nspire Handheld, because it is much more than just a calculator! Incidentally, the word "handheld" will be capitalized when prefaced by TI-Nspire, otherwise it will be lowercase.

- **Tool:** I routinely make reference to *tools* when talking about some of the features contained in the Graphs or Geometry application. When a tool is activated in either application, its associated icon is displayed in the upper-left corner of the screen. A tool remains active until you press either `esc` or `tab`, or when you begin using another tool. The Triangle tool is one such example. As the name implies, this tool allows you to draw a triangle.

   ✔ **TI-Nspire Computer Software:** Texas Instruments offers two types of software: TI-Nspire Student Software (which comes free with the purchase of a handheld) and TI-Nspire Teacher Software. Because these products are so similar, I often use the more generic term to describe both. Schools and our society are using computers more and more. With that in mind, one chapter on TI-Nspire Computer Software has been expanded to include three chapters in the update of this book.

Conventions Used in This Book

When I wrote this book, I had to train myself not to refer to the TI-Nspire unit as a *calculator*. This word is quite misleading, and it suggests that TI-Nspire has a limited amount of computing power. Rather, you will find that I refer to this product as a *device* or *handheld*.

As for pressing keys, I always refer to them by an icon represented by the physical key. For example, rather than saying "press the Enter key," I say "press [enter]." Sometimes, I refer to a sequence of keys to push, in which case I say "press [ctrl][⇧] to grab the object."

To access secondary functions, you must first press the [ctrl] key. I always tell you the exact keys to press to access such functions. For example, I say "press [ctrl][x²] to access the square root template."

The Touchpad (similar to a touchpad on a laptop) is located at the top of the keypad with the [⇧] key in the center. You also see small ◄► ▲ ▼ symbols located on the Touchpad. If I want you to move the cursor in a specific direction, say to the left, I tell you to "press the ◄ key repeatedly." If I simply want you to move the cursor to some other location, I say, "Use the Touchpad keys to move the cursor to a new location." If you want to move quickly, just swipe your finger across the Touchpad like you would on a laptop.

Foolish Assumptions

I assume that you are a beginning user who wants to discover the basics to get up and running with TI-Nspire. Why else would you choose to read a *For Dummies* book? Here are some other assumptions that I've made:

   ✔ You already own a handheld device or are planning on obtaining one soon.

   ✔ You are either an educator or a student. Being an educator myself, I found it tempting to write this book from a teacher's perspective. Although I do make some occasional references to teachers, you can expect that this book will work equally well for both teachers and students.

✔ As you see in Part IX, TI-Nspire Computer Software works nicely as a companion to the handheld device. I wrote these chapters under the assumption that you have some basic knowledge of how computers work. As you see in other sections of this book, a basic working knowledge of computers also comes in handy when working with your TI-Nspire Handheld (the *right-click* shortcut will become your best friend).

How This Book Is Organized

This book is organized around TI-Nspire's seven core applications. Because TI-Nspire applications often work together, it's hard to talk about them in isolation. However, I've done my best to write this book in such a way that you can jump in pretty much anywhere in the text without having to read the pages leading up to it. That being said, I recommend that you read this book sequentially to get the most out of it.

Part 1: Getting to Know Your TI-Nspire Handheld

In this part, I cover all the basics. This is where I introduce you to the philosophy behind TI-Nspire, the initial setup procedure, the document model, and all the tips and tricks that allow you to create, edit, and navigate documents quickly and efficiently.

If you are the type who likes to jump around from section to section, go right ahead. However, check out this part of the book first. It gives you the underlying structure to everything TI-Nspire.

Part II: The Calculator Application

This part gets into the first of seven core TI-Nspire applications. Here, you find out how to access a range of tools and commands that allow you to work with a variety of mathematical expressions and equations. In this part, I also start getting into how the Calculator application can "talk" to other applications. Finally, I introduce you to the computer algebra system of the TI-Nspire CAS Handheld.

Part III: The Graphs Application

The Graphs application represents one of TI-Nspire's most powerful applications. You find out how this application is used to provide a wide variety of different graph types, including functions, inequalities, scatter plots, polar equations, parametric equations, differential equations, and sequences.

I hope you'll also recognize the advantages that the Graphs application has to offer, providing a visual representation that can be analyzed right in the graphing window.

Part IV: The Geometry Application

The Geometry application provides one of TI-Nspire's most dynamic environments. Here, you find out how to work in an analytical environment, a plane geometry environment, or a combined analytic/plane geometry environment.

If you have some experience working with dynamic geometry software, you'll appreciate the smooth transition to this application. I hope you'll also recognize the advantages that the Geometry application has to offer, especially with its capability to have multiple representations on one page.

Part V: The Lists & Spreadsheet Application

Your experience with computer-based spreadsheet applications really pays off here. If words such as *fill down, cell,* and *formula* sound familiar, you'll have little trouble figuring out how to navigate this application. I also get into combining the Lists & Spreadsheet application with the Graphs, Geometry, or Data & Statistics application to perform regressions and investigate scatter plots.

Part VI: The Data & Statistics and Vernier DataQuest Applications

If you are working with the Lists & Spreadsheet application or the Calculator application, this application is perfectly suited for one- and two-variable analysis. In this part, you discover how to create and analyze a host of different statistical graphs, including dot plots, histograms, box plots, scatter plots, and summary plots. New color features allow incredible-looking comparative data representations. I also introduce Data Collection, a feature that works in conjunction with the Graphs, Geometry, Lists & Spreadsheet, and Data & Statistics applications.

The Vernier DataQuest application provides you with data collection tools that you have only dreamed of! Three views allow multiple representations of the data. Customize the data by selecting only the part of the data you would like to analyze. Using Lab Cradles, digital probes are now available. In addition, multiple probes are available using USB connections to a computer. If you are used to using EasyData on the TI-84, this application will blow you away!

Part VII: The Notes Application

The Notes application is the glue that holds together TI-Nspire's other applications. Simply put, this application makes the document model possible, eliminating the need to add paper to your activities as well as providing the continuity that makes your documents flow. You will find out how math expression boxes can become dynamic, linking interactively with the other applications.

Part VIII: TI-Nspire Computer Software

In this part, I talk about how TI-Nspire Computer Software makes a connection between your handheld device and your computer, allowing you to transfer files, take pictures of your handheld screen, back up your device, and upgrade the operating system.

TI-Nspire Computer Software allows you to quickly create and edit documents that are completely compatible with those that reside on your handheld device. In addition to providing the nuts and bolts of how to use this software, I give you several reasons why you might want to use it in the first place. I think you will agree that using the software to add a color image to the background of a Graphs page is bad to the bone!

Part IX: The Part of Tens

In Part IX, I give you a lot of good information — *quickly*. Here, I summarize ten great tips and shortcuts, periodically mentioned throughout the book, that are sure to save you lots of time. Finally, I resolve some common mistakes that I see in the classroom, and I equip you with the tools and knowledge to avoid the same pitfalls, as well as provide a way for you to access the vast array of resources that are available on the Internet.

Icons Used in This Book

This book uses four icons that help to emphasize a variety of points.

The text that follows this icon gives suggestions or shortcuts that help enhance your documents. These helpful little nuggets often pertain to the current material or suggest ways to extend or enhance the use of TI-Nspire.

The text that follows this icon tells you something that is truly worth remembering. I often use this icon to repeat something mentioned earlier in the book or to highlight information that will eliminate potential mistakes down the road.

The text associated with this icon is intended to warn you about more catastrophic mistakes, especially those that are difficult to troubleshoot. I'm thinking about that insidious issue that has no associated warning message. Nothing is more frustrating than dealing with an issue for which there appears to be no solution. This icon eliminates some of those issues.

I use this icon sparingly in this book. It gives you additional technical information that is intended only to satisfy your intellectual curiosity.

Where to Go from Here

This book is not the end-all. In fact, I periodically point you in the direction of some additional resources that are available to you. These resources include those provided with your TI-Nspire device when you purchased it as well as the abundance of resources found on TI's Web site, www.education. ti.com.

Regarding how to read this book, I mention earlier that you can read it sequentially or jump around as you see fit. If you are trying to locate something specific, refer to the table of contents or look it up in the index at the back of the book.

Many people try to memorize steps to accomplish a task on the TI-Nspire. I have even had a participant in one of my sessions who didn't want to pick up the TI-Nspire. She would have been happy to just take notes on all of the steps that we went through in the training. But is this the best way to learn new technologies? My suggestion to her (and you) is to experiment! Don't be afraid of making a mistake. Live dangerously. Grab different objects and observe the corresponding effects. If you don't like an effect, press ctrl esc to undo what you did. Fortunately, you don't have to take notes on the steps (they are already written down for you in this wonderful book). Don't be afraid to branch out and do additional explorations of your own.

Part I
Getting to Know Your TI-Nspire Handheld

The 5th Wave   By Rich Tennant

"You can sure do a lot with a TI–Nspire, but I never thought dressing up in G.I. Joe clothes and calling it your little desk commander would be one of them."

In this part . . .

This part gives you all the tools necessary to start creating and editing TI-Nspire documents. I encourage
you to start thinking of TI-Nspire as something more than
a calculator — something closer to a computer. From
installing the batteries to managing files to understanding
the document model, this part is sure to get you comfort-
able with the nuts and bolts behind TI-Nspire. I show
you how to use such time-saving shortcuts as right-click
and Ctrl+S.

Chapter 1

Using TI-Nspire for the First Time

*I*f you are brand new to TI-Nspire, I encourage you to start with this chapter. In this chapter, you begin to gain an appreciation of how TI-Nspire can help you understand mathematical concepts in a new way. You also find out about the different TI-Nspire products available and see some of the first steps to get up and running with TI-Nspire technology.

The Philosophy behind TI-Nspire

The best way to understand the philosophy behind TI-Nspire is to read this book and start playing with the device. However, let me whet your appetite now with a few thoughts about how TI-Nspire works and describe some things you can do with TI-Nspire that really showcase its capabilities.

Multiple representations

It has been demonstrated that students learn mathematical concepts more quickly and in greater depth when concepts are presented in multiple ways — that is, in algebraic, graphical, geometric, numeric, and verbal ways. TI-Nspire technology is all about multiple representations. In fact, TI-Nspire can display up to four different representations on a single screen.

Furthermore, these representations are dynamically linked. As you see in the next section, changes to one representation automatically affect the other representations, in real time, right on the screen. This highly interactive approach allows students to "see" the math, which enhances their ability to make mathematical connections and solve problems.

Figure 1-1 shows a simple example in which three representations of a concept are displayed. In the first screen in Figure 1-1, I give the algebraic representation of a given word description. In the second screen in Figure 1-1, I give the geometric representation and the numeric representation. Notice that the second screen contains two different applications on the same screen. With TI-Nspire, you have the option of displaying up to four different applications on one screen.

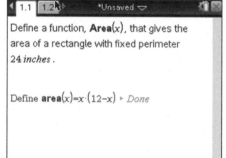 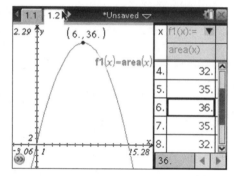

Figure 1-1:
Multiple
representa-
tions.

Linking representations

The idea of linking representations is another core feature that separates TI-Nspire from other calculators or handhelds.

Although it's nice to see multiple representations of a mathematical concept, it's really cool to have the option of manipulating one representation and watching the corresponding effect on another representation.

In the first screen in Figure 1-2, I change the size of a circle and watch the corresponding changes in *radius* and *area* measurements plotted on the coordinate plane in real time. The second screen in Figure 1-2 shows the radius and area data that automatically populates the Lists & Spreadsheet application as the circle is resized. This data represents the coordinates of each point that comprise the scatter plot.

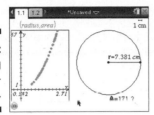 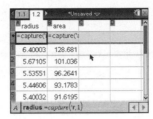 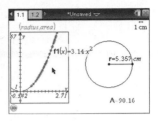

Figure 1-2:
Linking
representa-
tions.

Grab and move

The previous example helps to illustrate the grab-and-move philosophy inherent to TI-Nspire. You can also grab and move certain graphed functions and manipulate the axes themselves.

Imagine graphing $y = x^2$ in the previous example. TI-Nspire gives you the ability to grab the graph itself and change its shape. As you do this, the displayed equation on the screen updates automatically, again, in real time. Match the function to the scatter plot, and observe that the equation approximates $y = 3.14x^2$. Refer to the third screen in Figure 1-2.

The interactive feedback from this simple scenario allows students to explore and identify patterns and to make conjectures based on their observations. What a wonderful and interactive way to demonstrate the formula for the area of a circle!

The document model

In Chapter 2, you find out how TI-Nspire uses documents to engage students in interactive activities. A *document* is a TI-Nspire file that contains problems and pages. With TI-Nspire, you can create, edit, save, and review your documents using many of the same commands and file storage methods you use on a computer.

The document model provides students with three distinct advantages:

- ✔ Students can interact with the mathematics rather than just read about the mathematics in a static textbook.
- ✔ Students can pick up where they left off after leaving class.
- ✔ Students can work at home, either on their handhelds or on their computers.

As an educator, the document model provides you with these advantages:

✔ Teachers can prepare documents in advance and transmit them to students for use individually or in groups.

✔ Complicated constructions can be prepared in advance, thereby allowing students to focus on the math.

✔ Teachers can use multiple representations and the dynamic nature of TI-Nspire to really understand the underlying concepts behind the math.

The Computer Connection

If you are at all familiar with a PC, you should find the transition to TI-Nspire quite smooth. For starters, TI-Nspire documents consist of one or more pages, much like a document you might prepare using a word processor. As for working with your documents, you will find out about a variety of shortcuts that are virtually identical to those that you may already be using on your PC. For example, pressing the key sequence [ctrl][S] saves your work, pressing [tab] moves you to the next field in a dialog box, pressing [ctrl][menu] pulls up the context menu (the equivalent of a right-click menu on your computer), and so on. As for the right-click reference, get used to me talking about that feature. It's an incredible time-saver that you simply must take advantage of!

The more you remind yourself of this computer connection, the faster you will travel along the learning curve.

TI-Nspire versus TI-Nspire CAS

The TI-Nspire product line includes TI-Nspire and TI-Nspire CAS (both in the handheld and as a computer application). The TI-Nspire Handheld device performs numerical or *floating-point* calculations, much like those performed by the TI-83 and TI-84 product line. The TI-Nspire CAS Handheld has all the functionality of the TI-Nspire technology with two notable differences:

✔ TI-Nspire CAS technology has a built-in computer algebra system, which allows symbolic representation of numerical calculations — and the manipulation of algebraic expressions and processes (that is, you can expand binomials, find derivatives of algebraic expressions, and so on). For example, the solution to $x^2 = 12$ is given as

$x = -2\sqrt{3}$

or

$x = 2\sqrt{3}$

✔ TI-Nspire CAS Handheld does not include the snap-in TI-84 Plus Keypad (TI-Nspire CX does not have an interchangeable keypad either).

Because these devices have so much in common, this book can serve as a valuable resource for either handheld. I've included two chapters (Chapters 8 and 10) that specifically address some of the key features unique to TI-Nspire CAS. Throughout the book, I also occasionally point out some key differences between the two handhelds.

TI-Nspire versus TI-Nspire CX

The TI-Nspire product line has expanded once again! This time, color is the newest, coolest thing. Each of the seven applications has color display capabilities. Color is more than just a gimmick. The use of color on Data & Statistics pages makes it easy to compare one set of data with another. Multiple functions on the same Graphs page become much easier to distinguish when color is used. And, I cannot lie, I love accenting important terms with splashes of color on Notes pages.

The sleek new look of the TI-Nspire CX involves a slight rearrangement of the keys on the keypad. Compared to the TI-Nspire, the TI-Nspire CX has lost the bulk. TI-Nspire CX is more durable, thanks to a screen similar to what you may see on a touchscreen phone.

Figure 1-3 compares the keypads of the three generations of TI-Nspire Handhelds. You can see that with each new model has come a sleeker, more organized, and easier-to-see keypad. But, even if you have the TI-Nspire Clickpad, it is fully functional with the latest operating system from Texas Instruments.

Figure 1-3: Three generations of TI-Nspire Handhelds.

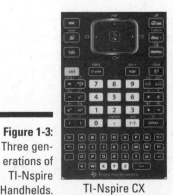 TI-Nspire CX

 TI-Nspire Touchpad

 TI-Nspire Clickpad

Installing Batteries

If you are like I am, you can't wait to finally open the box and start playing with your TI-Nspire Handheld. First, though, you must install the batteries, which are included when you purchase a TI-Nspire Handheld. Here's how to install AAA batteries in the TI-Nspire (Touchpad or Clickpad):

1. **On the back of the device, slide the tab to the right to release the keypad.**

2. **Slide the keypad down about ¼ inch and lift it out to reveal the battery compartment.**

3. **Insert the batteries, making sure that the + side of each battery faces toward the bottom of the device.**

4. **Place the keypad gently back in place, leaving about a ¼-inch gap at the top.**

5. **Slide the keypad up toward the display screen, applying enough pressure to snap it into place.**

An icon near the upper-right part of the screen indicates the battery status. When the battery status gets low, a small warning symbol appears. I usually wait to replace the battery until I see the warning symbol.

TI-Nspire CX and TI-Nspire Touchpad can be powered by a lithium battery. The TI-Nspire Rechargeable Battery is a lithium battery (similar to that for a cell phone), with each charge providing up to 100 hours of handheld power. The original TI-Nspire Clickpad handheld devices are not equipped to accept a TI-Nspire Rechargeable Battery. Here are the steps to install the battery:

1. **Use a small Phillips screwdriver to remove the two screws near the top of the back of the handheld.**

2. **Insert the TI-Nspire Rechargeable Battery, making sure to securely connect the white end of the battery wires to the handheld.**

3. **Screw the battery cover back in place.**

You can power your TI-Nspire Touchpad in three different ways. Similar to the way a hybrid car can use gas, electricity, or a combination of both, here are the different ways to power your TI-Nspire:

✓ The TI-Nspire Rechargeable Battery only

✓ The TI-Nspire Rechargeable Battery and four AAA alkaline batteries combined

✓ Four AAA alkaline batteries only

It can get really expensive to provide batteries for a classroom set of calculators (I know from personal experience). Wouldn't it be great to never have to buy batteries again?

TI-Nspire CX is powered solely by the TI-Nspire Rechargeable Battery. There are three convenient ways to charge the lithium battery:

- ✔ Use the adaptor to plug into a wall outlet.
- ✔ Use the cord (that came with your TI-Nspire purchase) to plug in to your computer's USB.
- ✔ Use the TI-Nspire Docking Station (see below).

Another innovative Texas Instruments product, the TI-Nspire Docking Station, can provide easy recharging, document transfer, and OS updates for a classroom set of handhelds. Both the TI-Nspire Rechargeable Battery and TI-Nspire Docking Station can be purchased at instructional dealers or at the TI online store.

Turning on the Unit and Going through the Initial Setup of TI-Nspire

To turn on your TI-Nspire device, press the ⌂on key.

After pressing ⌂on for the first time, you see a progress bar indicating that the operating system is loading. Eventually, you are greeted by a screen that prompts you to choose a language, giving you your first experience with a dialog box featuring drop-down menus.

To select the language, do the following:

1. **Press the center click key (🖱) to reveal the drop-down menu.**
2. **Use the ▲▼ keys on the Touchpad to highlight the language.**
3. **Press ⌨tab to highlight the OK button (as indicated by a dark outline) and press ⏎enter.**

If you are happy with the default settings of any dialog box, press ⏎enter and the settings take effect and close the dialog box at the same time. You don't need to tab through each field.

Next, you are prompted to select a font size. I happen to like the default medium font. However, if you want to change the font size, feel free to do so by following the same steps used for choosing a language.

Finally, you are greeted by a welcome screen, which describes some of the basic features of your TI-Nspire. Feel free to scroll through this document by pressing the ▼ key. Press ⌷enter⌷ to display the TI-Nspire Home menu.

The Three Zones of the TI-Nspire Keypad

The redesigned keypad is organized into three zones: Navigation, Math & Numeric Keys, and Alpha Keys. Keeping this in mind may help you as you get acquainted with TI-Nspire. A basic understanding of the TI-Nspire keypad helps you understand how to start navigating through documents quickly and efficiently.

The Touchpad and the ten keys near the top of the keypad make up the Navigation zone. The keys in the Navigation zone perform a variety of functions that you will find are quite similar to their computer counterparts. Here's a brief description of what each of these keys can do:

- ✔ ⌷esc⌷ **Escape:** This key removes menus or dialog boxes from the screen. For example, imagine that you have just activated the Perpendicular tool on a Graphs page. To remove this tool and activate the Pointer tool, just press the ⌷esc⌷ key.

- ✔ ⌷▥⌷ **Scratchpad:** Allows you to do calculations and graphs without having any effect on the document. Your calculations or graphs can be saved into existing documents.

- ✔ ⌷tab⌷ **Tab:** This key allows you to move to the next entry field in a dialog box. It also allows you to move around in certain applications. For example, pressing the ⌷tab⌷ key in the Graphs application moves you from the entry line to the work area. In the Lists & Spreadsheet application, the ⌷tab⌷ key moves you from one cell to the adjacent cell.

  Try pressing and holding ⌷⇧shift⌷ (Shift key) followed by the ⌷tab⌷ key. This key sequence moves you backward through a dialog box, just as it does on a computer.

- ✔ ⌷⌂on⌷ **Home:** This key displays the Home screen. The Home menu is where you can create a new document, access existing documents, and add pages to existing documents. It's also where you can adjust your system settings.

- ✔ ⌷doc▾⌷ **Documents:** This key activates the Documents management menu. Saving changes to documents, editing, and changing the page layout are just of few of the tasks that can be accomplished with this key.

✔ menu **Menu:** This key displays the menu associated with the current application (called the *application menu*). If you are on a Graphs page, you see one menu. If you are on a Lists & Spreadsheet page, you see a completely different menu.

Try pressing ctrl followed by the menu key. This key sequence acts just like a right-click on a computer mouse — it provides you with access to the *context menu*, a list of the specific options available based on the current cursor location or active object. This is the second time I've mentioned this feature and certainly not the last!

✔ ctrl **Control:** This key provides access to the secondary function or character located on a given key. For example, pressing ctrl ⌂on turns off your TI-Nspire Handheld.

✔ **Touchpad arrow keys:** In the middle of the Navigation zone of the keypad is the Touchpad (see Figure 1-4). If you look closely, you see the ▲ ▶ ▼ ◀ symbols clockwise from the top of the Touchpad. Simply put, pressing these keys allows you to move the cursor or pointer in any direction. Try swiping your finger lightly across the top of the Touchpad, which has the same effect as moving a mouse around on a mouse pad.

✔ 🖰 **Click:** Pressing this key selects objects on the screen, much like the Click button on your computer mouse. Press ctrl then 🖰 to grab objects. Alternatively, you can press and hold the 🖰 key momentarily to grab an object. This key will get a ton of use, because you have many clickable areas on TI-Nspire.

✔ del **Delete:** This key works just like the Backspace key on your computer. It deletes a single character of text or an entire selected object. Press ctrl del to clear the entire contents of a field.

✔ ⇧shift **Shift:** Pressing this key changes a lowercase letter to an uppercase letter. Pressing ctrl ⇧shift works like a Shift Lock key on a computer.

Figure 1-4:
Three zones
on the
TI-Nspire
keypad.

The Navigation
Zone

The Math & Numeric
Zone

The Alpha
Keys Zone

The Math & Numeric zone is centrally located on the keypad. The numeric keys are surrounded on both sides by math keys. These math keys have different functionality depending on whether you press the left or right part of the rocking keys. Many of these keys have secondary functions listed in small print above the keys.

The Alpha Keys zone is located near the bottom of the keypad. On each side of the alphabetical keys are special character keys. You may wonder why the keys are arranged in alphabetical order instead of like the keys on your computer. Most standardized tests require that calculators not have QWERTY (standard) keyboards.

Accessing menus and submenus

As I've already mentioned, the [menu] key gives you access to the menu options available in the current application. Pressing it once shows the top-level menu options. Some of these menu options have an arrow to their right, indicating that submenu options are available. To access a submenu, press ▼ to scroll down to the desired top-level menu option and then press ▶ to reveal the submenu. You may even find a third level of menu options, as shown in Figure 1-5.

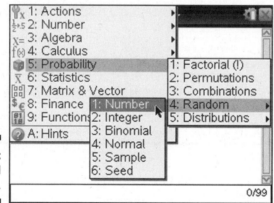

Figure 1-5:
Menus and
submenus.

To move back out of a series of submenus, just press [esc]. You need to press [esc] three times to remove the layers of menus shown in Figure 1-5.

Using the Scratchpad

The Scratchpad consists of two parts: Calculate and Graph. Pressing 🔳 once accesses one part, and pressing 🔳 again toggles to the other part. Alternatively, click (🔳) the tabs at the top of the Scratchpad to toggle from one to the other. The Scratchpad can also be accessed from the Home menu.

The Calculate part of the Scratchpad behaves exactly like a Calculator page, with the exception of not being able to access the Program Editor. See the first screen in Figure 1-6. The advantage of using the Scratchpad to do calculations is that it is available from anywhere! You don't even have to be in a document to access the Scratchpad.

The Graph part of the Scratchpad is a Graphs page that does not have any geometry functionality. My students love being able to do quick calculations or graphs by using the Scratchpad. Because the Scratchpad is completely separate from the document that is open, they won't have any record of their work (even if they save their document). However, you can always save your Scratchpad calculations and/or graphs by clicking (⌨) the ▼ arrow at the very top of the screen. Choose the Save to Document option to have a record of your work (in either a new document or the current document). See the second and third screens in Figure 1-6.

TIP

If your Scratchpad gets too cluttered, you can clear the Scratchpad. Click (⌨) the ▼ arrow at the top of the screen and choose Clear Scratchpad. Both the Calculate and Graph parts of the Scratchpad will be cleared.

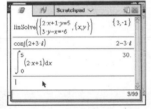 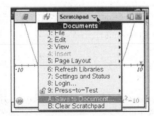 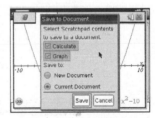

Figure 1-6:
Using the
Scratchpad.

Configuring the Settings

Okay, now that you know a bit about the TI-Nspire keypad, it's time to configure the settings. To access the settings, press ⌂on⇨Settings. The first screen in Figure 1-7 shows six choices. Choosing Handheld Setup is a good place to start. The second and third screens in Figure 1-7 show some of the resulting choices you have. If you want to conserve battery life, consider changing the Power Standby field (which turns off your handheld after the specified time of inactivity), the Hibernate field (similar to a computer, it will take an extra-long time to turn on your handheld because you must reboot the OS), and the Auto Dim field (the screen will not turn off, but will dim slightly to save power after a specified time of inactivity). To make a change to the Handheld Setup screen, use the Touchpad arrow ▶ to expand the selection and click (⌨) to make your selection. I recommend selecting (⌨) the Enable Tapping to Click check box, which allows you to tap the Touchpad instead of clicking it to make a selection. To make your changes effective, press ⌨ tab (or ▼) to navigate to the OK button and then press 📱 or ⌨ enter.

 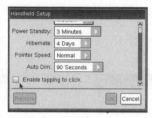 

Pressing 🏠on⇨Settings⇨Status shows you the version of the operating system currently running on your handheld. You will also see a battery status for up to three types of batteries: AAA Batteries, Handheld Rechargeable Battery, and Cradle Rechargeable Battery (only used with TI-Nspire Navigator).

To access the Document Settings screen on your handheld, press 🏠on⇨Settings⇨Document Settings. See the first screen in Figure 1-8. I strongly recommend pressing Make Default when making changes to these settings. When you do, the following prompt will appear: `Apply your settings to open document and save them as default for new documents and Scratchpad?` Click (📱) OK to set the same settings to your Scratchpad. Trust me, it could get really confusing if your calculator is set to Radian and your Scratchpad is set to Degree!

Pressing the Restore button defaults to the original TI-Nspire factory settings.

Now that the Calculator settings have been changed to your liking, it is time to change the Graphs application settings. You can access the Graphs & Geometry settings directly from a Graphs page. Press ⌨ctrl⌨I⇨Add Graphs⇨Menu⇨Settings. See the second screen in Figure 1-8. Changing the Display Digits field affects the precision of the points that are found in the Graphs environment. Notice that the Graphing Angle field (a graphed function) and the Geometry Angle field (a geometric construction) are considered different settings.

Four check boxes can be selected on the Graphs & Geometry Settings screen. See the third screen in Figure 1-8. Here is an explanation of each:

✓ **Automatically Hide Plot Labels:** The default setting on TI-Nspire is to label all functions that are graphed in a Graphs page. If you would like to disable this feature, click (📱) to select this box.

✓ **Show Axis End Values:** The end values on a Graphs page are the x and y maximum and minimum values. If you would not like for these to appear on the ends of the x and y axes, click (📱) to deselect the box.

✔ **Show Tool Tips for Function Manipulation:** I strongly recommend not selecting this box. If you do (and you have been warned), a message will appear every time you grab and move an object.

✔ **Automatically Find Points of Interest:** You will love this feature! On a Graphs page, this feature can help you find the zeros, maximums, minimums, and so on.

 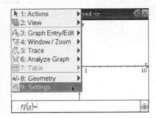 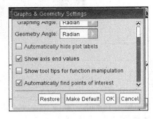

Figure 1-8: The Graphs & Geometry Settings screen.

Always choose Make Default so that you can change the settings on the Scratchpad to match the settings of the Graphs page.

Switching Keypads

The TI-Nspire (Touchpad or Clickpad) can operate with two different keypads (not at the same time): the TI-Nspire keypad and a TI-84 Plus Keypad. When the TI-Nspire keypad is installed, you are using TI-Nspire. When the TI-84 Plus Keypad is installed, your handheld device works exactly like any TI-84 Plus Silver Edition device. This means that you actually have two handheld devices!

If a TI-84 Plus Keypad was not included with the purchase of your TI-Nspire, you may be able to acquire one at no additional cost. Log on to `education.ti.com/84keypad` and complete the online form.

To change the current keypad, follow these steps:

1. **On the back of the device, slide the tab to the right to release the keypad.**

2. **Slide the keypad down about ¼ inch and lift it out to reveal the battery compartment.**

3. **Place the new keypad gently in place, leaving about a ¼-inch gap at the top.**

4. **Slide the keypad up toward the display screen, applying enough pressure to snap it into place.**

Each time that you change keypads and turn on the unit, you must wait for the new operating system to load (as indicated by a progress bar).

Updating the operating system on your TI-Nspire automatically updates the operating system on the TI-84.

I work only with the TI-Nspire keypad in this book. If you are interested in finding out more about the TI-84, refer to *TI-84 Plus Graphing Calculator For Dummies,* by C. C. Edwards (published by Wiley).

Chapter 2

Understanding the Document Structure

· ·

In This Chapter

▶ Using the application menu and Documents menu in My Documents

▶ Opening, closing, and saving files

▶ Understanding the document structure

▶ The seven core TI-Nspire applications

▶ Using three different views: Full Page, Page Sorter, and My Documents

· ·

*I*n this chapter, I give you the information you need to better understand the file management system inherent in TI-Nspire. I then show you how you can open an existing document, and I give you the tools to move around an open document quickly and efficiently.

Navigating the Home Menu

The Home menu automatically displays each time you turn on the TI-Nspire and can also be accessed by pressing ⌂on. In many ways, the Home menu is the starting point for the many activities that are possible on the TI-Nspire.

Some of the options available on the Home menu are the following: calculating and graphing using the Scratchpad, adding applications to documents, changing the settings, and managing documents. Use the arrow keys on the Touchpad to navigate the options and press ⎙ or enter to select the option that is highlighted. Alternatively, press the letter or number that corresponds to the option you would like to choose.

The Home menu contains five choices that help with document management:

- **New Document:** Opens a new document.

- **My Documents:** Opens the file browser. Just like a computer, this is where all the files and folders are stored.

- **Recent:** Accesses a list of the five most recently opened documents.

- **Current:** Takes you to the document that is currently open.

- **Settings:** Gives you options for changing the settings on your handheld.

Managing Files and Folders

Computers allow you to save files to folders. TI-Nspire does, too. The My Documents view allows you to save and organize your files and folders.

The My Documents view

My Documents allows you to view all the folders and files on your TI-Nspire Handheld. To open the My Documents view, press 🏠on⇨My Documents.

The first screen in Figure 2-1 shows the different folders I have available in My Documents. To view the contents of any folder, use the Touchpad keys to highlight the folder and press the 🔘 key to expand the folder (or press ▶). Repeat this process if you want to collapse a folder (or press ◀). In the second screen, I opened the Algebra 2 folder.

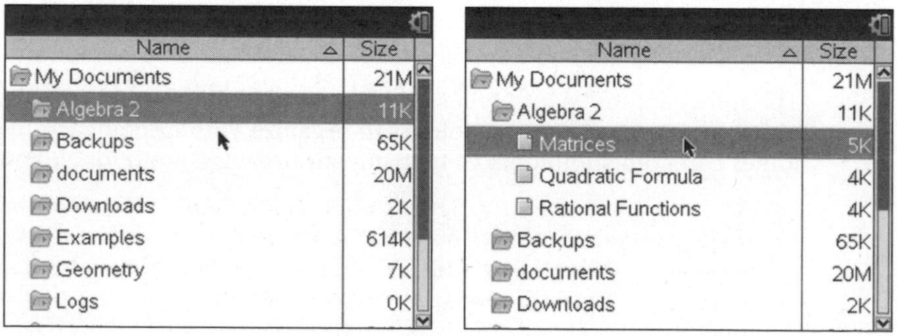

Figure 2-1:
The My
Documents
view.

Name	△	Size
📁 My Documents		21M
📁 Algebra 2		11K
📁 Backups		65K
📁 documents		20M
📁 Downloads		2K
📁 Examples		614K
📁 Geometry		7K
📁 Logs		0K

Name	△	Size
📁 My Documents		21M
📁 Algebra 2		11K
📄 Matrices		5K
📄 Quadratic Formula		4K
📄 Rational Functions		4K
📁 Backups		65K
📁 documents		20M
📁 Downloads		2K

The My Documents application menu

The My Documents menu differs slightly depending on whether you have a folder or file highlighted. To access the My Documents menu, press the menu key. Figure 2-2 shows the menu options available for a highlighted folder (first screen) and a highlighted file (second screen).

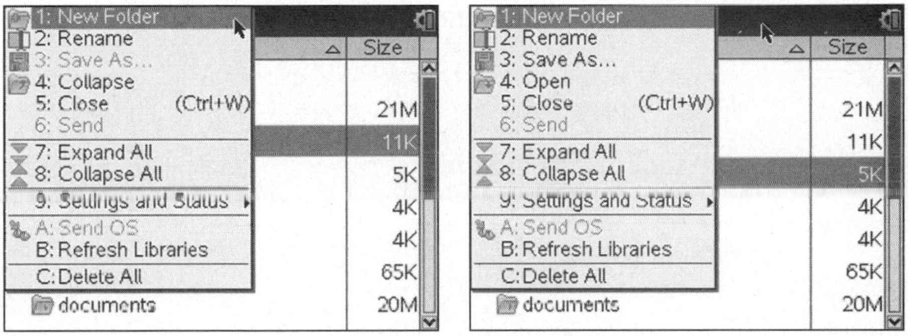

Figure 2-2:
The My
Documents
menu.

Notice that the My Documents menu contains an alternate way to expand or collapse folders. Two other related options, Expand All and Collapse All, are available on the My Documents menu (press menu).

Using folders to organize files

The My Documents menu can be used to help you organize your folders and files. Look back at Figure 2-1 and notice that I have named two folders by subject, Algebra 2 and Geometry. To add another folder, say, Precalculus, press ⌂⇧⇨My Documents, then press menu⇨New Folder. A new folder, temporarily named Folder1, appears. Using the alpha keys, type the name of your new folder and press enter.

TI-Nspire can hold a lot of files, so it's a good idea to spend time thinking about how you want to use folders to organize your documents. In fact, each folder can hold subfolders to help in your organization of documents.

Renaming files and folders

Perhaps you made a mistake naming a file or folder. Simply highlight the file or folder and press menu⇨Rename. A secondary box appears around the highlighted file or folder with the existing name highlighted in gray. Start typing the new name and press enter when you are finished.

Filenames and folder names can be 255 characters long. You can use almost any character, including spaces, as part of a filename or folder name.

Other My Documents menu items

You may have noticed a few other options located on the My Documents menu. Two of these options, Save As and Send, are only available if you have a file highlighted rather than a folder. I describe how to use the Send option to send a file to another handheld in Chapter 4. I also talk about how to use Send OS in Chapter 4.

You also have the options to Save As and Open a highlighted file. Selecting the Save As option automatically opens a dialog box that allows you to type a new filename, then you can save within the same folder or a different folder. This feature saves a copy and preserves the original file. See Figure 2-3.

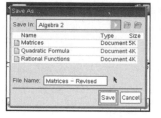  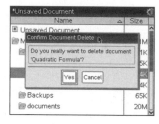

Figure 2-3: Using the Save As command.

To delete a document, right-click, ctrl menu ⇨Delete (or press del over a highlighted document). As a safeguard, a prompt asks whether you really want to delete the file. See the second and third screens in Figure 2-3.

The Documents menu

The Documents menu, accessed by pressing doc▾, is the one menu that is available no matter where you are in TI-Nspire. Furthermore, the Documents menu options never change. For example, you can access the Documents menu from a Graphs page or from within the My Documents view and you see the same menu choices. Keep in mind, though, that certain menu items may not be available for use, as indicated by the light-gray font.

The first screen in Figure 2-4 shows the top-level Documents options. The second and third screens show the secondary Documents options for File and Edit, respectively.

Shortcut keystrokes are embedded in the menus of the TI-Nspire. Notice that the shortcut for the Save command, (Ctrl+S), is written directly to the right of the command on the menu. See the second screen in Figure 2-4.

Figure 2-4:
The
Documents
menu
accessed
from the My
Documents
view.

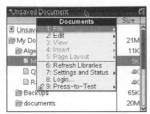 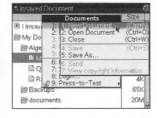 

Notice that the Edit menu offers several editing options similar to those found on a computer. Additionally, several of these options can be accessed by using keyboard shortcuts, the same ones that are used on a PC.

Here is a complete list of the Documents menu options that have corresponding keyboard shortcuts:

- ✔ [doc▾]⇨File⇨New Document (Shortcut: [ctrl] [N])
- ✔ [doc▾]⇨File⇨Open (Shortcut: [ctrl] [O])
- ✔ [doc▾]⇨File⇨Close (Shortcut: [ctrl] [W])
- ✔ [doc▾]⇨File⇨Save (Shortcut: [ctrl] [S])
- ✔ [doc▾]⇨Edit⇨Undo (Shortcut: [ctrl] [Z] or [ctrl] [esc])
- ✔ [doc▾]⇨Edit⇨Redo (Shortcut: [ctrl] [Y])
- ✔ [doc▾]⇨Edit⇨Cut (Shortcut: [ctrl] [X])
- ✔ [doc▾]⇨Edit⇨Copy (Shortcut: [ctrl] [C])
- ✔ [doc▾]⇨Edit⇨Paste (Shortcut: [ctrl] [V])

The right-click menu

Let me officially introduce the *right-click option*. When you perform a right-click, you pull up a contextual menu or shortcut menu that gives a list of available options depending on the application that is running, the objects that are currently selected, or the cursor location.

To access the contextual menu on a TI-Nspire Handheld, press ⌐ctrl⌐ ⌐menu⌐.

Within the My Documents view, the contextual menu includes a collection of options that reside in the My Documents application menu and on the Documents menu. See Figure 2-5.

One main advantage of using the contextual menu is that you can avoid digging through menus and submenus to find a specific feature or function.

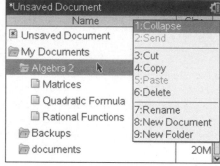
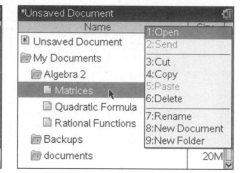

Figure 2-5: The contextual menu associated with the My Documents view.

Documents, Problems, and Pages

Now that you've had an opportunity to find out about the file management system of TI-Nspire, it's time to start looking at the files themselves.

Opening, closing, and saving files

To open a file, you must be in the My Documents view. Simply highlight a specific file and press ⌐enter⌐ or ⌐·⌐. You can also highlight a file and press ⌐menu⌐➪Open (or ⌐ctrl⌐⌐O⌐). The first page of the new document is displayed. To go back to the My Documents view, press ⌐on⌐➪My Documents.

In the My Documents view, open files are designated by an asterisk symbol (*) to the left of the filename.

You have many ways to close a file on the TI-Nspire. Press ⌐doc⌐➪File➪Close (or press ⌐ctrl⌐⌐W⌐). Alternatively, click (⌐·⌐) the X in the upper-right corner of the screen. On the TI-Nspire Handheld, you cannot have two files open simultaneously. Therefore, one way to close a file is to open another file. If you made any changes to a file, you are prompted to save the currently open file before the new file opens.

As for saving files, here's how it works. A new document is not saved — it resides in TI-Nspire's local memory, just like a new (but unsaved) document on a computer. To save a new, unnamed document, follow these steps:

1. Press `doc▾`⤵File⤵Save to open the Save As dialog box.

2. By default, the cursor is located in the File Name field. Type the filename.

3. To specify a different folder location, press `⇧shift` `tab` to move up to the Save In field. Press `⬚` to reveal and select an available folder or press `⇧shift` `tab` again and `⬚` to create a new folder and type the folder name.

4. At any time, press `enter` to put your choices into effect and close the dialog box. Alternatively, press `tab` until the OK or Cancel button is highlighted and press `enter`.

To save a previously saved document with the current name, press `ctrl` `S` (or press `doc▾`⤵File⤵Save).

To save an open file under a different name (thus preserving the original file), press `doc▾`⤵File⤵Save As.

Understanding how documents are structured

Every time you use TI-Nspire, you are either working on an existing document or working with a new document. Again, I remind you that this experience is very similar to your experience working on a computer, especially when using a word processor. However, significant differences exist, too, and I discuss those differences in the following sections.

The seven core applications

TI-Nspire has seven applications from which to choose. Here's a list of the applications and a brief description of what each does.

✔ **Calculator application:** In this application, you perform calculations. You also enter and view expressions, equations, and formulas, all of which are displayed in a format similar to what you see in a textbook. A variety of built-in templates is also available to give you the power to represent just about any mathematical concept symbolically.

✔ **Graphs application:** In this application, you graph equations, expressions, and a variety of functions. Variables and sliders allow you to investigate the effect of certain parameters dynamically. Analyze the graph to find critical points and the values of local extrema.

✔ **Geometry application:** In this application, you can explore synthetic geometry concepts, that is, geometry not associated with the coordinate plane. Also, the Geometry application allows you to integrate coordinate geometry and synthetic geometry. Watch as connections between these two areas are made dynamically, in real time.

✔ **Lists & Spreadsheet application:** In this application, you investigate numeric data, some of which is captured from the Graphs application and some of which resides entirely within the Lists & Spreadsheet application. Like a computer spreadsheet program, this application allows you to label columns, insert formulas into cells, and perform a wide range of statistical analyses.

✔ **Data & Statistics application:** Used in conjunction with Lists & Spreadsheet, this application allows you to visualize one-variable and two-variable data sets. Data & Statistics allows you to create a variety of statistical graphs, including scatter plots, histograms, box-and-whisker plots, dot plots, regression equations, and normal distributions. You can also manipulate a data set (either numerically or graphically) and watch the corresponding change in the other representation.

✔ **Notes application:** The Notes application enables you to put math into writing. Three templates make the Notes application a robust and integral part of any TI-Nspire document. With the Notes application, you can pose questions, review or write geometric proofs, and provide directions for an activity. Interactive math boxes link to all the other applications.

✔ **Vernier DataQuest application:** This application can be used along with probes (like the CBR2 motion detector) to collect real data. There are three views available within the DataQuest application that allow for multiple representations of the data. You can even discard the parts of the data that you do not want to include.

Dividing documents into problems and pages

Figure 2-6 shows the first three pages of a document titled *00GettingStarted*. Each page has a tab associated with it, and a white tab indicates that the page is active. Inactive pages are designated by a gray tab. A small ▶ symbol located to the right of the last tab indicates that additional pages comprise the entire document. As for the numbering system itself, the first number indicates the problem number and the second number indicates the page number.

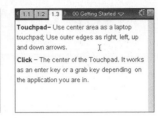

Figure 2-6:
Documents,
problems,
and pages.

Press ⟨ctrl⟩ ▶ to move to the next page and press ⟨ctrl⟩ ◀ to move to a previous page. Alternatively, click (🖉) the tabs or the small ◀▶ symbols located to the right or left of the tabs near the top of the page. (Refer to Figure 2-6.)

Understanding how variables interact

The seven core TI-Nspire applications do not reside in isolation. In fact, most documents consist of a variety of applications, all working in conjunction with one another. Therefore, variables defined on one page are available for use on another page, as long as the pages are part of the same problem.

Figure 2-7 helps illustrate this point. In the first screen of Figure 2-7, I have graphed the function $f1(x) = x^2$ in problem 1, page 1. I used a split screen and typed **f1(3)** on a Calculator page and pressed ⟨enter⟩. The result is 9, because the function $f1(x)$ is the same function defined within the same problem.

Perhaps I want to use $f1(x)$ again but do not want to change its original definition. To accomplish this, I insert a new problem. In the second screen in Figure 2-7, I have defined $f1(x) = x^3$. Because this page is part of problem 2, you can be assured that $f1(x)$ is still equal to x^2 in problem 1. I can use the same variables for different purposes, if they are in different problems.

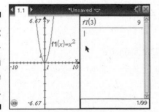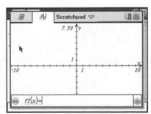
Figure 2-7:
Shared variables within the same problem.

I have accessed the Scratchpad in the third screen in Figure 2-7. Notice that the defined function $f1(x)$ is empty! Variables and functions defined in a document do not interact with the Scratchpad; the Scratchpad is completely separate from your document.

Choose Your Level

TI-Nspire offers three different ways to view your documents:

- ✔ Full Page view
- ✔ Page Sorter view
- ✔ My Documents view

Each view has certain advantages, and when used in combination, these views can allow you to complete a variety of tasks quickly and efficiently.

Full Page view

In the earlier section "Dividing documents into problems and pages," I talk about how documents are comprised of problems and pages. The Full Page view allows you to see one complete page at a time. The screens shown in Figure 2-7 are all examples of pages shown in the Full Page view.

Dealing with more than one application on a page

With TI-Nspire, you can view up to four applications on a single page. Figure 2-8 shows one such case. In the first screen in Figure 2-8, notice the dark box located around the Lists & Spreadsheet application found in the lower-left corner of the screen, indicating that this application is currently active.

To move to other applications on the same page, press ⌷ctrl⌷⌷tab⌷. The second screen in Figure 2-8 shows the result of this action: The upper-left application is now active. In general, pressing ⌷ctrl⌷⌷tab⌷ always moves you to the next application in a clockwise direction. Just keep pressing ⌷ctrl⌷⌷tab⌷ until you reach the desired application.

My favorite way to move from one application to another is to swipe my finger across the Touchpad to activate the cursor. Then, move the cursor to the desired application and click (⌷) to activate.

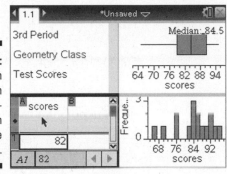
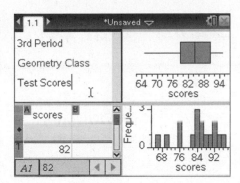

Figure 2-8:
Moving from application to application within the same page.

Page Sorter view

The Page Sorter view gives you a bird's-eye view of an open document (similar to a thumbnail view on a computer). To enter the Page Sorter view, press ⌃ ▲ from within the Full Page view. Figure 2-9 shows the Page Sorter view of the *00GettingStarted* file. Using the Touchpad keys, you can highlight any problem or page. If you highlight a problem number, you can press 🔲 to expand/collapse the pages within the problem.

Moving around from page to page and from problem to problem

Move to any page and press ⏎ to bring up the page in Full Page view. In the second screen in Figure 2-9, I have highlighted problem 2, page 3. The third screen in Figure 2-9 shows this page in Full Page view after pressing ⏎.

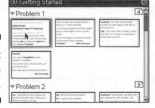
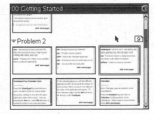
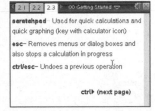

Figure 2-9:
The Page Sorter view.

Changing the problem or page order

The Page Sorter view offers a convenient way to change the order of problems and pages within a document. To accomplish this, highlight a page and press ⌃ 🔲 to grab the page. Use the Touchpad keys to move the page, and press ⏎ to drop it in place. See Figure 2-10.

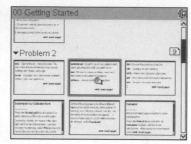

Figure 2-10:
Using the
Page Sorter
view to
change the
page order.

The My Documents view revisited

Toward the beginning of this chapter, I talk about how you can access the My Documents view by pressing ⌂on⇨My Documents. Well, it turns out that you can access the My Documents view a second way.

If you are in Full Page view, press ctrl ▲ to move to the Page Sorter view. Press ctrl ▲ again to access the My Documents view. If you are already in the Page Sorter view, just press ctrl ▲ once to access the My Documents view. You will see that the current open filename is highlighted. Press enter to jump back to Full Page view. If you choose to open a different file from within the My Documents view, you will be prompted to save the current file (assuming that you have made changes to the file).

Chapter 3

Creating and Editing Documents

In This Chapter

▶ Creating documents consisting of pages and problems

▶ Working with Page Layout to configure screens with two or more applications

▶ Grouping and ungrouping apps

▶ Using Cut, Copy, and Paste within different TI-Nspire views

▶ Undoing and redoing your actions

*I*n this chapter, I give you the tools you need to create and edit documents. You find out how to insert pages and problems, customize the page layout, and take advantage of TI-Nspire's convenient editing features to ensure that your documents look and act their best.

Creating a New Document

So you're finally getting up the nerve to create a new document. After turning on your TI-Nspire, you're likely to find that you're already in an open document. Here are three ways to open a new document:

✔ **Press [doc▾]⇨File⇨New Document.** If you currently have a document open, you may be prompted to save the current file.

If you're asked, "Do you want to save *<Document Name>?*" be careful! The question refers to the previous document, not the new document you're opening! You have three choices: Press [enter] to save the document, press [tab][enter] to open the new file without saving the current document, or press [tab][tab][enter] to cancel the transaction. See the first screen in Figure 3-1.

✔ **Press [⌂ on] and select New Document from the given options.**

✔ **Press [ctrl][N] (shortcut for New Document).**

After dealing with the prompt to save the current open document, you see the second screen shown in Figure 3-1. Use the Touchpad arrow keys to select an application and then press [enter]. A new page based on the application you choose opens. The first page of the document is denoted problem 1, page 1.

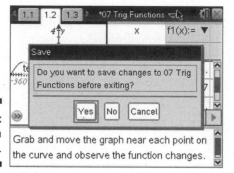
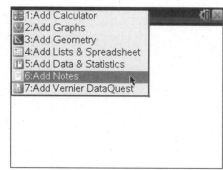

Figure 3-1: Opening a new file.

Adding pages to your documents

In this case, I added a Notes page (the first screen in Figure 3-2 shows that I typed some instructions that refer to the next page). To insert a Graphs page (choose this application to graph a function), do one of the following:

- ✔ Press [ctrl][doc▾], use the Touchpad to highlight Add Graphs from the available options, and press [enter]. (See the second screen in Figure 3-2.)

- ✔ Press [⌂ on] and choose the Graphs icon.

- ✔ Press [ctrl][I] (shortcut for Insert Page) and select Add Graphs.

The tabs at the top of each page indicate the problem number and the page number. The third screen in Figure 3-2 shows problem 1, page 2.

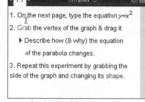
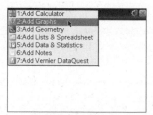
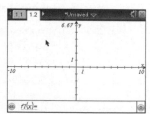

Figure 3-2: Adding a page.

A single Notes page   Select Add Graphs   Blank Graphs page

Adding problems to your document

A new problem allows you to use the same variables as another problem without creating conflicts. A new problem can be added by pressing ⌜doc▾⌝⇨Insert⇨Problem. Alternatively, a new problem can be inserted by pressing ⌜ctrl⌝▲ and then ⌜menu⌝⇨Insert Problem.

After adding a new problem, you're greeted by a screen prompting you to pick an application, just as when you add a new page.

Saving your work

Very soon after you create a new document, it's a good idea to save the file. The quickest and easiest way to save your new file is to press ⌜ctrl⌝⌜S⌝ (shortcut for ⌜doc▾⌝⇨File⇨Save). Because this file wasn't saved previously, a dialog box opens, giving you complete control over the filename and the folder in which to save.

Follow these steps to change the folder location or to create a new folder.

1. **Press ⌜doc▾⌝⇨File⇨Save or the shortcut key sequence ⌜ctrl⌝⌜S⌝.**

2. **Type the name of the file in the File Name field.**

3. **Press ⌜⇧shift⌝⌜tab⌝ to move from the File Name field to the folder that's currently open, or just move your cursor and press ⌜🖰⌝.**

   All the folders contained in the My Documents menu are listed.

4. **Highlight the name of the folder you wish to save your document to and press ⌜enter⌝ to select the folder.**

   Navigating a dialog box on the TI-Nspire is very similar to navigation on a computer. Pressing ⌜tab⌝ advances to the next field and pressing ⌜⇧shift⌝⌜tab⌝ goes back to the previous field. Alternatively, because a dialog box is a clickable area, you can navigate by moving your cursor and pressing ⌜🖰⌝ to choose the field that you wish to edit (like moving your mouse on a computer).

5. **To create a new folder, press ⌜⇧shift⌝⌜tab⌝ to highlight the new folder icon and then press ⌜enter⌝ to create a new folder; type the name of the folder in the highlighted field.**

6. **Press ⌜enter⌝ to accept file saving information and close the dialog box.**

   Alternatively, press ⌜tab⌝ until OK is highlighted, and then press ⌜enter⌝ or ⌜🖰⌝.

If you don't want to change the folder location, complete Steps 1, 2, and 6. As you continue to work on your document, you can periodically save your work by pressing ⌜ctrl⌝⌜S⌝.

Configuring a Page Layout

TI-Nspire allows you to display up to four applications on one screen. Of course, you need to balance your desire to display several different representations of a problem with a need to keep the screen uncluttered.

Configuring pages with up to four applications

Say you want to solve the equation $x^2 - x - 1 = 0$ using the zero feature on a Graphs page and then confirm your answer using the Numerical Solve command in the Calculator application.

In the first screen in Figure 3-3, you can find the positive solution to this equation by using the Graph Trace tool (more about how to do this in Chapter 9). To add a Calculator page to the same screen, press doc▾⇨Page Layout⇨Select Layout. In the second screen in Figure 3-3, notice several layout options available. For this example, I chose Layout 2, which brings up the third screen in Figure 3-3.

Figure 3-3:
Changing
the page
layout.

 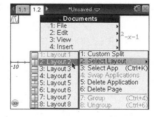 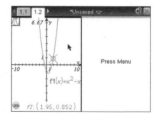

Notice the dark box surrounding the Graphs application in the third screen in Figure 3-3, indicating it's the current active application. Use the Touchpad to move your cursor to the right side of the screen and then press 🔄 to activate the new application (alternatively, press ctrl tab). In the first screen in Figure 3-4, notice that the dark box is now on the right side of the screen. Press menu ⇨Add Calculator.

In the second and third screens in Figure 3-4, press menu , select Add Calculator, and open the Numerical Solve command to verify the graphical solution to $x^2 - x - 1 = 0$. To access the Numerical Solve command from the Calculator application, press menu ⇨Algebra⇨Numerical Solve. The complete expression used in the last screen in Figure 3-4 is nSolve(f1(x)=0,x,1,3). This syntax indicates that you want to look for the solution to $x^2 - x - 1 = 0$ on the interval from $x = 1$ to $x = 3$.

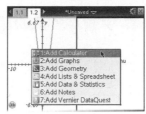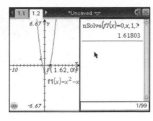

Figure 3-4: Switching applications.

Creating a custom split

In the first screen in Figure 3-5, I added a third application — Lists & Spreadsheet — to the current screen. To add this third application, press [doc▾]⇨Page Layout⇨Select Layout⇨Layout 7. The goal here is to generate a rather famous sequence to see how it might relate to the solution to the equation $x^2 - x - 1 = 0$. To generate the sequence, I pressed [menu]⇨Data⇨Generate Sequence and configured the dialog box as shown in the second screen in Figure 3-5. As you can see, it's hard to see the values in the spreadsheet because there's so little room. Use the Custom Split tool to see whether you can improve upon the space utilization. Here's how:

1. **Press [doc▾]⇨Page Layout⇨Custom Split to activate the Custom Split tool.**

2. **Press the arrow keys on the Touchpad to adjust the vertical split down the middle of the page and the horizontal split between the two applications on the right side.**

3. **Press [enter] when you're happy with the split to exit the Custom Split tool. See the third screen in Figure 3-5.**

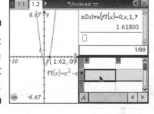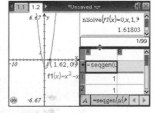

Figure 3-5: Using the Custom Split tool.

You must press [enter] after configuring the custom split. If you press [esc] after configuring the split, your changes don't take effect and the Custom Split tool closes.

Using the Swap Applications tool

Although the Custom Split tool in the preceding section helped a bit with the layout, it'd be even better to swap the position of the Graphs application (located on the left side of the screen) with the Lists & Spreadsheet application (located on the bottom-right corner of the screen). Fortunately, the Swap Applications tool located in the Page Layout menu accomplishes this task. Follow these steps:

1. **Press [ctrl] [tab] until one of the applications you want to swap is active (as indicated by the dark box surrounding the application).**

   In the first screen in Figure 3-6, I activated the Graphs application.

2. **Press [doc▾]⇨Page Layout⇨Swap Applications to activate the Swap Application tool.**

   Notice that the Graphs application pulses and the ⟨⟩ icon is located in the Calculator application (see the first screen in Figure 3-6). Pressing [enter] swaps these two applications. Keep in mind, though, that you want to swap the Graphs and Lists & Spreadsheet applications.

3. **Use the arrow keys on the Touchpad to move the ⟨⟩ icon to the Lists & Spreadsheet application (see the second screen in Figure 3-6) and then press [enter] to make the swap with the Graphs application.**

   The third screen in Figure 3-6 shows the result of this change (after a bit of additional cleanup). Much better!

Figure 3-6: Using the Swap Applications tool.

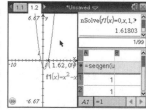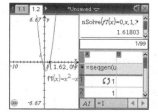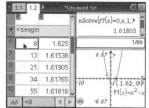

Notice in the third screen in Figure 3-6, the ratio of consecutive terms of the famous Fibonacci sequence has a value (the Golden Ratio), which is approximately equal to the solution to $x^2 - x - 1 = 0$!

Changing a page layout

Say you're an indecisive person and decide that you really don't want the Calculator page cluttering your screen. Therefore, you want to go back to a layout that includes only two applications — the Graphs application and the Lists & Spreadsheet application.

Follow these steps to accomplish this task:

1. **Move your cursor to the Calculator application and press ⏹ to select the application.**

   Alternatively, press [ctrl][tab] until the Calculator application is active.

   See the first screen in Figure 3-7.

2. **Press [doc▾]⇨Page Layout⇨Delete Application to delete the Calculator application.**

   Notice that the Calculator application is removed, and the page layout is back to a split screen. In fact, the Graphs application is back on the left side of the screen. (See the second screen in Figure 3-7.) You can always use the Swap Application tool again if you really want the Graphs page on the right side of the screen.

 The Documents menu can be accessed by pressing [doc▾] or by clicking on the down arrow (▾) to the right of the document name at the top center of the screen. (See the third screen in Figure 3-7.) The down arrow (▾) is just one of many clickable areas on the TI-Nspire.

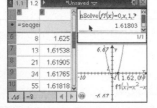 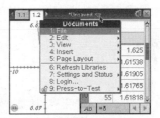

Figure 3-7: Changing the page layout.

Grouping and Ungrouping Applications

What if you change your mind again? If you decide you don't want both applications on the same page, you can use the Ungroup command to separate the split page into two separate pages.

The Group and Ungroup commands can be accessed from two views — Page Sorter view and Full Page view. To change the view to Page Sorter, press [ctrl]▲. Use the Touchpad to highlight the page that you want to ungroup. Press [menu]⇨Ungroup to separate the split page into two pages (or press [ctrl][6]). (See the first screen in Figure 3-8.)

If you decide that you'd rather view the applications on a split page after all, you can use the Group command to group applications on two consecutive pages. Before you access the Group command, turn to the first page that you want to group. To access the Group command from Full Page view, use the Documents menu by pressing [doc▾]⇨Page Layout⇨Group (or press [ctrl][4]).

(See the second screen in Figure 3-8.) This action automatically groups the next page into the page where the command was initiated. This action can be repeated but is limited to four applications on one page.

If you paid close attention to the menus in the first and third screens in Figure 3-8, you may have noticed the shortcuts (ctrl 4 and ctrl 6) are listed right after the commands. TI-Nspire includes the shortcut key sequences in the menu if a shortcut exists. What a great way to help you remember the shortcut key sequences and learn new ones as well!

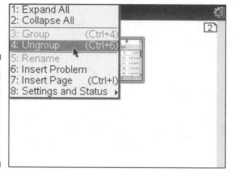
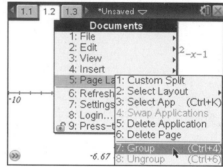

Figure 3-8:
Ungrouping
and
grouping
applica-
tions.

Managing Documents: Cut, Copy, and Paste

A number of options are available to quickly edit or manipulate existing documents. Some of these options I mention in Chapter 2. In this section, I talk briefly about how Cut, Copy, and Paste can be used in each of the three document views.

In My Documents view

You can access My Documents view in two ways — by pressing 🏠on⇨My Documents or by pressing ctrl ▲ twice from Full Page view (or once from Page Sorter view).

Perhaps you have a file that's already been created. You want to keep that file as is but also want to copy it and edit it for a slightly different purpose. From within My Documents view, just highlight the file and press ctrl C (shortcut for copy) and ctrl V (shortcut for paste). A second file appears with the same name preceded by *Copy of.* You can then highlight the filename and press menu⇨Rename to change the name to whatever suits you. Be careful, this technique won't work with a file that is open.

You can cut (⌊ctrl⌋⌊X⌋), copy (⌊ctrl⌋⌊C⌋), or paste (⌊ctrl⌋⌊V⌋) any unopened file or folder from within My Documents view.

In Page Sorter view

Page Sorter view is accessed by pressing ⌊ctrl⌋ ▲ once from Full Page view. Alternatively, you can press ⌊doc▾⌋⇨View⇨Page Sorter. However, I think this second option is more time-consuming and not worth the effort.

From within Page Sorter view, you can also use Cut, Copy, and Paste to perform a variety of edits. For example, you can cut (or copy) any page by highlighting the page and pressing ⌊ctrl⌋⌊X⌋ (or ⌊ctrl⌋⌊C⌋ to copy). When a page has been cut (or copied), it can be pasted back into the same problem or to another problem. You can even paste a cut (or copied) page into another document assuming it's the last thing you cut (or copied).

If you copy a page into another problem that uses the same variables, you are greeted by the message Cannot add to a problem because one or more variables of the same name already exist and cannot be overwritten.

To delete an entire problem from within Page Sorter view, highlight the problem number with the Touchpad arrow keys and then press ⌊del⌋. After you delete a problem, all subsequent problems are renumbered.

Rather than copy only one page at a time, you can save time by copying the entire problem. To accomplish this, follow these steps:

1. **Access Page Sorter view by pressing** ⌊ctrl⌋ ▲ **once from Full Page view.**

2. **In Page Sorter view, use the Touchpad to highlight the name of the problem.**

   See the first screen in Figure 3-9.

3. **Press** ⌊ctrl⌋⌊C⌋ **(or press** ⌊menu⌋⇨Edit⇨Copy**) to copy the problem.**

4. **In Page Sorter view, highlight the location where you want to copy your problem.**

5. **Press** ⌊ctrl⌋⌊V⌋ **to paste the copied problem.**

From within a page

Cut, Copy, and Paste work the same way from within Full Page view. To cut or copy text, a mathematical expression, and so on, hold down the Shift key (⌊⇧shift⌋) and press the Touchpad arrow keys until you select what you want. Press ⌊ctrl⌋⌊X⌋ to cut the selected region (or ⌊ctrl⌋⌊C⌋ to copy) and save it to the

Clipboard. Then use Paste ([ctrl][V]) either in the same page or a different page. The second and third screens in Figure 3-9 show the result of copying and pasting text.

Figure 3-9: Using Copy and Paste with problems and pages.

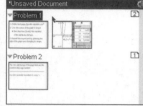 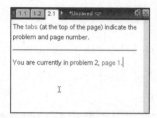 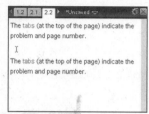

Pressing [ctrl][K] from within Full Page view, followed by [ctrl][C] and [ctrl][V], splits the page vertically and inserts a copy of the entire page side by side with the original page.

Perhaps you want to copy an entire page and paste the copy to a new page, all from within Full Page view. To accomplish this, follow these steps:

1. **Press [ctrl][K] from within Full Page view to select the entire page.**

2. **Press [ctrl][I] to insert a new page.**

3. **Press [esc] to remove the prompt for choosing an application, revealing *Press Menu* in the center of the screen.**

4. **Press [ctrl][V] to paste the copied page.**

The Undo and Redo Commands

We all make mistakes once in a while. Sometimes these mistakes seem catastrophic, especially if you just cut or deleted a large quantity of text, entire pages, or a complete document. Not to worry! To undo a series of changes, use the Undo command, denoted by ↷, by pressing [ctrl][esc] repeatedly until you have restored your work. You can also access the Undo command by pressing [doc▾]⇨Edit⇨Undo or the shortcut key sequence [ctrl][Z], that is also used on a computer.

The Redo command enables you to move forward through a series of commands that were just undone. Press [doc▾]⇨Edit⇨Redo, or use the shortcut [ctrl][Y] to access the Redo command.

Keep these two options in mind at all times. They work everywhere and allow you to eliminate mistakes in a flash. This tool is very useful for someone who changes their mind often. Wouldn't it be nice if life had an Undo command!?

Chapter 4

Linking Handhelds

. . . .

. . . .

In this chapter, I tell you how to communicate between two TI-Nspire Handhelds. In Part VIII, I tell you how to communicate between your TI-Nspire Handheld and your computer.

Sending and Receiving Files or Folders

Your TI-Nspire Handheld comes with two USB cables. One cable has a standard USB connector on one end and a small Mini-A USB connector on the other end. This cable is used to communicate between the TI-Nspire Handheld and the computer.

The other cable has the small Mini-A USB connector on both ends. I call this the *unit to unit cable,* and I use it to transfer files between two TI-Nspire Handhelds. To connect two handhelds, follow these steps:

1. **Insert one end of the unit-to-unit cable into either handheld and press firmly to establish the connection.**

2. **Insert the other end of the same cable into the second handheld. Press firmly.**

To send a document or folder, follow these steps on the sending handheld:

1. **Press ⌂on⇨My Documents to enter the My Documents view.**

2. **Highlight the document or folder you want to send and then press menu⇨Send.**

The file transfer begins automatically. When the process is complete, you see screens similar to those shown in the first two screens in Figure 4-1.

The third screen in Figure 4-1 shows what happens if the receiving calculator already contains a file with the same name — TI-Nspire appends a number after the filename and doesn't overwrite the original file.

Figure 4-1:
Transferring
documents
between
two hand-
helds.

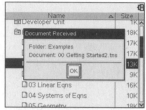

Sending handheld screen   Receiving handheld message   Dealing with a duplicate
                                                                    filename

I always put the sending handheld on the left and the receiving handheld on the right. That way, I never forget which handheld does the sending.

No action is required on the part of the receiving handheld. In fact, it doesn't even need to be powered on; it turns on automatically when the unit-to-unit cable is attached.

The sending handheld always puts a sent document in a folder on the receiving handheld that has the same name as the sender's folder. If no such folder exists on the receiving handheld, the sender creates a folder with this name.

Here are some other rules that are followed when sending documents from one handheld to another:

- **The maximum length for a document name is 255 characters.** If the document already exists with the same name on the receiving handheld, the filename is truncated to allow for renaming the file.

- **All variables associated with a sent file are included in the transferred file.**

- **If a problem arises (usually because the cables aren't pressed in fully), the transmission times out after 30 seconds.**

Sending Your Operating System to Another Handheld

It's always a good idea to periodically check whether you have the latest operating system (OS). If you do, you can take advantage of any new features that TI has come up with for your TI-Nspire handheld.

In this section, I tell you how to transfer the TI-Nspire operating system from one handheld to another. For starters, connect the two handheld devices with the unit-to-unit cable, as described earlier in this chapter. Then follow these steps on the sending handheld:

1. **Press [⌂ on]⇨My Documents to enter My Documents view.**

2. **Press [menu]⇨Send OS.**

The file transfer automatically begins, as indicated by the message `Sending OS <version number>. Do not unplug cable.` on the sending handheld. Additionally, you see a progress bar on the sending handheld. Keep in mind, the transfer may take several minutes to complete. You don't see a progress bar (or any action for that matter) on the receiving handheld.

After the transfer completes, you see messages on both handhelds. The sending handheld displays the message, as shown in Figure 4-2. The receiving handheld shows a progress bar indicating that the new OS is being installed.

Figure 4-2:
Transferring
an OS
between
two hand-
helds.

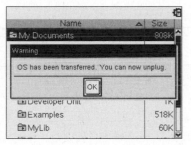

After the updated OS is installed on the handheld, you're taken through the same initial setup screens (that is, Choose Language and Font Size), as described in Chapter 1. I usually just press [enter] three times to complete the initial setup.

The automatic power-down feature of TI-Nspire is disabled during transfers. Therefore, it's a good idea to check your battery level before starting an OS transfer by pressing [⌂ on]⇨Settings⇨Status.

Part II

The Calculator Application

In this part . . .

This part takes a look at the first of seven core TI-Nspire applications. As the name implies, I show you how to enter and evaluate numerical expressions. I also show you how to access functions and commands on the Calculator application menu as well as the Catalog. Additionally, I cover how variables are defined and used in the Calculator application and discuss how this concept helps the Calculator application "talk" to other TI-Nspire applications.

One chapter is dedicated to TI-Nspire CAS, the powerful handheld device with the built-in computer algebra system. Specifically, I talk about how CAS functionality looks and acts in the Calculator application.

Chapter 5

Entering and Evaluating Expressions

In This Chapter

▶ Working with primary and secondary keys

▶ Dealing with results in different forms

▶ Working with expressions and answers

▶ Using the history and last answer to work efficiently in the Calculator application

▶ Accessing commands, symbols, and templates from the Catalog

▶ Understanding the Calculator application menu

In this chapter, I show you how to use the Calculator application to do what its name implies — perform a wide variety of calculations. However, the name *Calculator application* is somewhat of a misnomer. As you see in subsequent chapters in this part of the book, the Calculator application can communicate with all other applications and perform a variety of tasks that go well beyond the basics.

Evaluating Expressions Using Primary and Secondary Keys

In this section, I show you how to evaluate mathematical expressions directly from the *primary* keys (as defined by the functions or characters located directly on the keys themselves) and the *secondary* keys (as defined by the colored functions or characters located toward the top of some keys). For example, x^2 is the primary key for squaring an expression and the secondary key for taking the square root ($\sqrt{}$) of an expression. To access a secondary key, press [ctrl] followed by the primary key.

To add a Calculator page, press [ctrl][I]⇨Add Calculator.

Start with a very simple expression, entered and evaluated by pressing only primary keys. In the first screen in Figure 5-1, I typed ③④③÷⑤ to find the value of 3 + 3 ÷ 5. The result of this calculation, found by pressing ⏎, is shown in the second screen in Figure 5-1. Notice that two things happen after you press ⏎. The original typed expression is shown with a stacked fraction, and the result is also expressed as a stacked fraction. When possible, TI-Nspire displays expressions and results in *pretty print* — that is, in the format that you typically see in a math textbook or other print source.

But why force the TI-Nspire to convert your expression to *pretty print?* Typing the secondary key for division (ctrl ÷) gives you a Fraction template. This template allows you to type intense math problems (like the complex fraction in the third screen in Figure 5-1) with minimal use of parentheses. Students who use this template avoid the parentheses errors that used to be very common in my math classroom because the expression they type on the Calculator page looks exactly like the problem in their textbook.

As for the fractional result, TI-Nspire attempts to display all rational values as fractions. In the second screen in Figure 5-1, I typed ③ ctrl x² ①② to find the value of 3 times the square root of 12. Notice that the result is expressed as a decimal, which happens any time a result is an *irrational* number (that is, when the decimal portion neither terminates nor repeats and thus can't be written in fractional form).

When using TI-Nspire CAS, the expression

$$3\sqrt{12}$$

returns the exact value

$$6\sqrt{3}$$

See Chapter 8 for more information about the symbolic representation of results associated with TI-Nspire CAS.

This result is shown to four decimal places. To change the number of displayed digits, press ⌂on⇨Settings⇨Document Settings. A dialog box opens and the first field, Display Digits, allows for changing the number of displayed digits.

Pressing OK affects only the settings within the current document (and doesn't change the Scratchpad settings). I strongly recommend choosing Make Default every time you make a change so that the Scratchpad settings change as well as future documents that you create.

Figure 5-1:
Evaluating
expressions
with primary
and second-
ary keys.

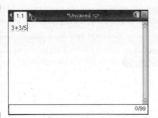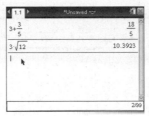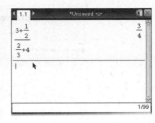

Consider that you want to evaluate $\log_5 25$. The logarithm function is a secondary key located on the ⌈10ˣ⌉ key. The first screen in Figure 5-2 shows that
a Logarithm template appears after pressing ⌈ctrl⌉⌈10ˣ⌉. Notice the two small
dashed rectangular (fields) with the cursor located in the leftmost field. The
first field defines the base of the logarithm, and the second field gives the
value for which you want to evaluate this logarithm. Type ⌈5⌉ to specify the
base. To move to the next box, press ⌈tab⌉ (or ▸). Type ⌈2⌉⌈5⌉ and press ⌈enter⌉ to
complete the calculation, as shown in the second screen in Figure 5-2.

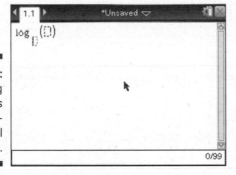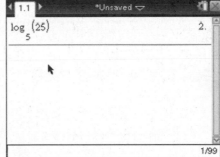

Figure 5-2:
Working
with fields
in a math-
ematical
expression.

 Always use the ⌈tab⌉ key to move from field to field in a mathematical expression just like you do on a computer. Pressing ⌈⇧shift⌉⌈tab⌉ moves your cursor to the
previous field.

 If you leave the first field in the Logarithm template blank, TI-Nspire uses a
default base of 10.

Dealing with Very Large and Very Small Results

true, however, when using TI-Nspire. Take, for example, the calculation $\boxed{(}\boxed{\text{ctrl}}$ $\boxed{\div}\boxed{5}\boxed{\blacktriangledown}\boxed{2}\boxed{\blacktriangleright}\boxed{)}\boxed{\wedge}\boxed{7}\boxed{4}$ (five-halves raised to the 74th power). The first screen in Figure 5-3 shows the exact result as a stacked fraction in excruciating detail! This result is roughly equivalent to 2.8×10^{29}.

Now try evaluating two-fifths raised to the 74th power (press $\boxed{(}\boxed{\text{ctrl}}$ $\boxed{\div}\boxed{2}\boxed{\blacktriangledown}\boxed{5}\boxed{\blacktriangleright}\boxed{)}\boxed{\wedge}\boxed{7}\boxed{4}$). This extremely small result, approximately equal to 3.57×10^{-30}, also displays as a stacked fraction in all its glory. See the second screen in Figure 5-3.

Very large or very small results often don't fit on a single screen. To view an entire result, press ▲ once to highlight the answer. Press $\boxed{\text{enter}}$ and then use the ◀▶ keys to scroll through and view the entire answer.

So just how large an exponent can you raise five-halves? Try raising this number to the 1419th power. TI-Nspire displays the answer as a stacked fraction. Now try an exponent of 1420. This time, the result displays in scientific notation. Now try raising five-halves to the 2513th power. TI-Nspire can't handle this calculation and displays an Error Overflow message. (See the third screen in Figure 5-3.) TI-Nspire CAS deals with overflow issues by displaying the result as ±∞ and displaying the message `Overflow replaced by ∞` or `-∞` at the bottom of the page.

Figure 5-3: Very large and very small results.

  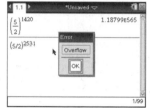

Controlling the Form of a Result

Clearly, you're sometimes better off seeing an answer expressed as a decimal or in scientific notation. Here are four ways to force an approximate result, ranked in order from easiest to hardest:

- ✔ **Press $\boxed{\text{ctrl}}\boxed{\text{enter}}$ to evaluate an expression.** This action approximates the value of the expression by accessing a secondary function represented by the symbol ≈.

- ✔ **Include a decimal point somewhere in your calculation.** In Figure 5-4, I typed $\boxed{3}\boxed{+}\boxed{3}\boxed{\div}\boxed{5}\boxed{.}$ to evaluate the expression in decimal form.

- ✔ **Use the Convert to Decimal command.** Type $\boxed{3}\boxed{+}\boxed{3}\boxed{\div}\boxed{5}$ and then press $\boxed{\text{menu}}$⇨Number⇨Convert to Decimal. This command can also be found in the Catalog, which I talk about later in this chapter.

✔ **Use the Approx command to force a result in decimal form or scientific notation.** This command can be typed using the alpha keys. It can also be found in the Catalog.

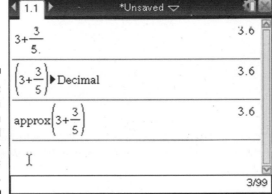

Figure 5-4:
Forcing a
result in
decimal
form or
scientific
notation.

Using History

TI-Nspire offers a convenient way to copy and paste an expression in order to perform similar calculations.

Accessing previously used commands

Consider that you want to use the quadratic formula to solve the equation $x^2 + 3x - 9 = 0$. For starters, access the Fraction template, a secondary function located on the ÷ key, which pastes the Fraction template to the entry line with a blank numerator field and a blank denominator field. Type the numerator; then press tab to move to the denominator, type 2, and press enter to evaluate the expression. See the first screen in Figure 5-5.

The second solution to this equation can be evaluated by making a slight edit to the expression just entered. Here are the steps to follow:

1. **Press ▲ twice to highlight the previous expression (see the second screen in Figure 5-5).**

2. **Press enter to paste this expression to the entry line.**

3. **Use the keys on the Touchpad to move the cursor to the right of the + sign in the numerator of the expression.**

4. **Press del to clear the + sign and press ⊟ to insert a subtraction sign.**

5. **Press [enter] to evaluate this revised expression (see the third screen in Figure 5-5).**

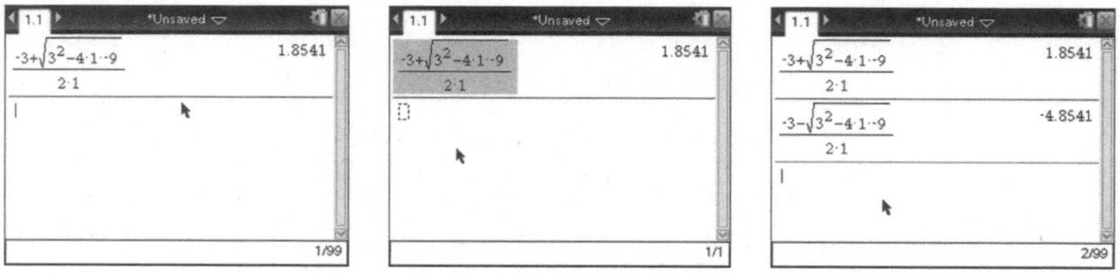

Figure 5-5: Accessing and evaluating previously used expressions.

You can fill a total of 99 lines on a single Calculator page. If you don't clear your history (see the next section), any of up to 98 previous entries can be pasted to the entry line.

Clearing the Calculator history

To clear the Calculator history, simply press [menu]➪Actions➪Clear History.

If you choose to clear the history, all previously defined variables and functions retain their current values. Use the Undo feature ([ctrl][esc]) to restore the history if you mistakenly delete it.

Using the Last Answer

Each time you perform a calculation, the result is stored to TI-Nspire's local memory as the Ans variable. As a result, you can quickly access this stored variable and use it in subsequent calculations.

Consider, for example, that you want to teach students how to evaluate an algebraic expression, such as $2x^2 + 1$ for $x = -8$. This problem can be easily evaluated by entering the entire expression once on the entry line. However, perhaps you want to take students through the process, step by step, to place additional emphasis on the order of operations. Here's how the Last Answer feature can accomplish this task (see Figure 5-6):

1. **Type –8 on the entry line and press [enter].**

   This stores the number –8 to the last answer Ans variable.

2. **Press the [x²] key.**

   This action automatically pastes the Ans variable to the entry line and raises it to the second power. (See the first screen in Figure 5-6.)

3. **Press [enter] to square the last answer.**

   This stores the number 64 to the variable Ans. (See the second screen in Figure 5-6.)

4. **Press [×][2] to multiply the last answer by 2.**

   The entry line reads Ans•2.

5. **Press [enter] to evaluate this expression.**

   This stores the number 128 to the Ans variable.

6. **Press [–][1] to paste the expression Ans–1 to the entry line. Press [enter] to find that the value of $2x^2 - 1$ for $x = -8$ is equal to 127 (see the last screen in Figure 5-6).**

Figure 5-6:
Using the
last answer
variable,
Ans.

The last answer variable, Ans, is automatically pasted to the entry line if, after evaluating an expression, you press [x²], [÷], [×], [–], or [+].

Notice that whenever the Ans variable appears in the entry line, it is replaced with a numeric value after you press [enter].

Using the last answer variable in an expression

The last answer variable can also be accessed by pressing [ctrl][(-)], which is very helpful if you want to take your last answer from a calculation and use it one or more times in a subsequent calculation. For example, consider a cylinder of known volume and height (volume = 50 and height = 10); you want to find the volume of a sphere with the same radius. Using the formula for volume, you can solve for the radius, as shown in the first screen in Figure 5-7. The last answer from this calculation, accessed by pressing [ctrl][(-)], is used as the radius to find the volume of the sphere (see first and second screens in Figure 5-7).

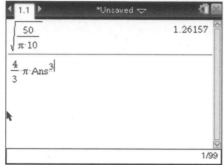
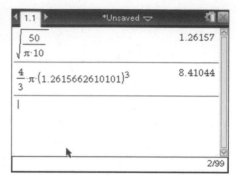

Figure 5-7:
Using [ctrl] [(−)]
to access
the last
answer
variable.

Using the last answer variable to generate a sequence

You can also use the last answer variable to generate a sequence, such as 2, 5, 14, 41, . . . Each term of this sequence is found by multiplying the previous value by 3 and subtracting 1. Follow these steps to generate successive terms of this sequence:

1. **Press [2] [enter].**

   This action stores the first value of the sequence, 2, to the last answer variable.

2. **Press [×][3][−][1] to paste the expression Ans•3−1 to the entry line.**

3. **Press [enter] to evaluate this expression.**

   Notice that the expression Ans•3−1 is replaced by 2•3−1, indicating that the value of Ans in the previous step is equal to 2.

4. **Continue to press [enter] to generate additional terms of this sequence (see the first screen in Figure 5-8).** The second and third screens in Figure 5-8 show how to generate a more complicated sequence of numbers. Perhaps you recognize the numerators and denominators of each fractional answer as terms from the Fibonacci sequence.

Figure 5-8:
Using the
last answer
variable to
generate a
sequence.

Copying, Pasting, and Editing Expressions and Answers

You may have noticed that pressing the ▲ key alternatively highlights previous answers and expressions from the calculator history. When a previous answer or expression is highlighted, you can press ⌷ctrl⌷⌷C⌷ to copy the answer or expression to local memory. You can then use ⌷ctrl⌷⌷V⌷ to paste this copied item as often as you want in future calculations.

For example, perhaps you're using the distance formula to find the distance between two points. After entering and evaluating the expression for the first time, press ▲ twice to highlight the original expression, press ⌷ctrl⌷⌷C⌷, press ▼ twice to move back down to the entry line, and press ⌷ctrl⌷⌷V⌷ to paste the expression to the entry line. You can then use the Touchpad arrow keys and the ⌷del⌷ key to edit the expression. See Figure 5-9.

Figure 5-9:
Copying and pasting expressions.

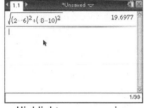 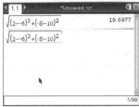 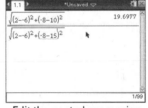

Highlight an expression   Paste the expression   Edit the pasted expression

You can also highlight an expression and simply press ⌷enter⌷ to paste it to the entry line. However, the method described in this section allows you to press ⌷ctrl⌷⌷V⌷ to paste the expression as often as desired for subsequent calculations.

You can also copy just a part of an expression from the Calculator history. To do this, follow these steps:

1. **Use the ▲ key to highlight a previous expression.**

2. **Press ◀ or ▶ to move the cursor to the left or right of the part of the expression you want to highlight.**

3. **Press and hold ⌷⇧shift⌷, and then use the ◀ or ▶ key to drag the highlighting to the left or right.**

4. **Press ⌷ctrl⌷⌷C⌷ to copy the highlighted item.**

Using the Catalog (📖)

The Catalog, accessed by pressing the 📖 key, contains a list of all system functions, commands, symbols, and Expression templates. These items can be pasted to the entry line of the Calculator application.

Moving from category to category

Press the 📖 key to open the Catalog. Within the Catalog, five categories are indicated by the numbered tabs located at the top of the screen. Here's a brief description of what each category represents:

1: 📖

2: ∫∑

3: ∝β

4: 📖

5: 📚

- ✓ **Category 1:** Contains a list of all commands, functions, and symbols, in alphabetical order. See the first screen in Figure 5-10.
- ✓ **Category 2:** Contains a list of all math functions, grouped by topic. See the second screen in Figure 5-10.
- ✓ **Category 3 (the Symbol palette):** Contains a list of all mathematical symbols. See the third screen in Figure 5-10.
- ✓ **Category 4 (the Expression templates):** Contains a list of Expression templates including square root, piecewise functions, and matrices.
- ✓ **Category 5:** Shows a list of all Public Library objects.

Figure 5-10: The Catalog.

 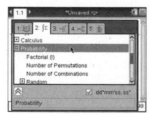 

To move from one category to the next, just type the number associated with each category. For example, press ④ to access the Expression templates. Alternatively, the five tabs are clickable areas if you move your cursor and press 🔘.

Accessing functions and commands

When you're in a desired category, use the Touchpad keys to highlight your choice and press [enter] to paste it to the entry line. Keep in mind that the items contained in the second tab are grouped by topic. To expand a topic, such as Probability, highlight the topic name and press 🔘.

The alphabetical list contained in the first category of the Catalog is quite extensive. Fortunately, a few tricks allow you to access specific commands quickly. If you know the name of a command, press the letter corresponding to the first letter of the command. For example, to access the Rand command, press ⓇR to jump to those commands beginning with R. Then use the Touchpad to scroll down to the Rand command. When your command has been highlighted, press 〔enter〕 to paste it to the entry line. No matter what tab is active in the Catalog, press 〔ctrl〕〔3〕 to page-down through the entire list or 〔ctrl〕〔9〕 to page-up through the list. To jump to the last item in the Catalog, press 〔ctrl〕〔1〕 (the equivalent to the End key on your computer); to jump to the first item in the Catalog, press 〔ctrl〕〔7〕 (the equivalent of the Home key on your computer).

Determining syntax

Notice on the first screen in Figure 5-11 that the bottom of the screen shows the syntax associated with the rounding command, Round. To see an expanded view of this syntax (or any command or symbol in the Catalog), press 〔tab〕 to highlight the bottom portion of the screen and then press 〔enter〕 to expand this area (see the second screen in Figure 5-11). Depending on the command that you have highlighted, you may see several rows of syntax.

Any part of the syntax that's contained in brackets is optional. For example, the Round command, by default, rounds up to the number of display digits specified in the Document Settings. The optional second number in the argument determines the number of decimal places to include in the result. The third screen in Figure 5-11 shows the result of Round(12.365) and Round(12.365,2).

You may have noticed the check box next to Wizards On in Figure 5-11. To select the Wizards On check box, press 〔tab〕 until the word *Wizards* becomes highlighted and press 🛈. Then press 〔tab〕 and 〔enter〕 to execute the command that is highlighted. When the Wizards On check box is selected, certain functions (such as the LinRegMx command) open with a dialog box to help you enter arguments in an expression.

Figure 5-11:
Using syntax help to execute a command.

Each of the areas in the Catalog is clickable. Just swipe the Touchpad to discover your cursor and feel free to click on a box or tab as an alternative to pressing 〔tab〕 repeatedly.

Alternative ways to access the Symbol palette and Expression templates

The third and fourth categories can also be accessed directly. To access the Symbol palette, press the secondary keys [ctrl][⊞]. To access the Expression templates, press [⊞].

Evaluating expressions in multiple ways

TI-Nspire's primary and secondary keys, application menu, and Catalog provide for a lot of power when working in the Calculator application. The combination of all the keys and menus also means you often have many ways to complete a task.

Earlier, you evaluated the expression $\log_5 25$. Turns out, you can evaluate this expression several ways. Here's a list of these methods, in no particular order:

- Press [ctrl][10ˣ] to access the Logarithm template from a secondary key.
- Press [⊞] to access the Expression templates, highlight the Logarithm template, and press [enter] to paste it to the entry line.
- Press [⊞][4] to access the Expression templates through the Catalog.
- Press [⊞][1] to access the alphabetical list of functions and commands and then follow these steps:

    a. *Press [L] to jump down to those functions and commands that begin with L.*

    b. *Use the Touchpad keys to highlight Log (and press [enter] to paste the command to the entry line).*

    c. *Complete the expression by typing **log(25,5)** and pressing [enter].*

- Type [L][O][G] using the alpha keys to open the same command found in the Catalog, and then follow step *c* to complete the expression and find its value.

Clearly, some of these options are more efficient than others. However, those who like choices should be quite happy. The many options also demonstrate the versatility of TI-Nspire.

Exploring the Calculator Application Menu

Each application in TI-Nspire has its own unique application menu, accessed by pressing the [menu] key. Some top-level applications (the first menu choices

REMEMBER

you see after pressing [menu]) have submenus as indicated by the small ▶ located to the right of the menu choice. To view a submenu, highlight the menu choice using the Touchpad and press [enter] or 🔲.

Press [esc] or ◀ to back out of a submenu.

The Calculator application menu has so many different tools that I simply can't provide a sufficient description of each tool. Rather, I give you a sampling of some of my favorite tools from each top-level menu choice.

Working with Number tools

Notice that the top-level menu choices are grouped by topic. For example, press [menu]➪Number to access a variety of choices for working with numbers, including fractions, decimals, prime factorization, greatest common divisor, and so on. Figure 5-12 provides a brief sampling of some of the tools found in the Number submenu.

Figure 5-12: The Number submenu commands.

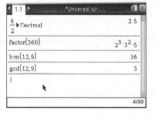
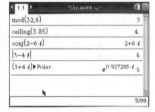

TIP

If you don't know the syntax of a function or command, you can look it up in the reference material provided with your TI-Nspire device, download a manual from www.education.ti.com, or access the command from the Catalog, which also provides help with the proper syntax.

Using the powerful Algebra tools

Press [menu]➪Algebra to access the Algebra submenu. The Numerical Solve command solves one-variable equations with ease. In the Catalog, the syntax for the Numerical Solve command is as follows:

```
nSolve(Equation, Var[=Guess], lowBound, upBound)
```

See the various ways that I used this command to solve an equation in the first screen in Figure 5-13. If you don't specify the interval in which to look for a solution, TI-Nspire returns the value that's closest to the origin. Using the Guess feature or specifying the interval helps you find the second solution of a quadratic equation. But wouldn't it be nice to find both solutions to a quadratic equation using just one command?

TI-Nspire CAS uses a Solve command that generates multiple solutions for given polynomial equations.

The Polynomial Root Finder command can find multiple roots to polynomial equations with just one command. To open the tool, press [menu]⇨Algebra⇨Polynomial Tools⇨Find Roots of Polynomial. An easy-to-use wizard shows a dialog box to gather the necessary information and correctly enter the syntax into the Polynomial Root Finder command. See the result in the second screen in Figure 5-13.

The Simultaneous Linear Equations Solver is another powerful tool in the Algebra submenu. To open the tool, press [menu]⇨Algebra⇨Solve Systems of Linear Equations. Let the wizard guide you, and enter the system of equations that you want to solve. See the third screen in Figure 5-13.

Both the Polynomial Root Finder and the Simultaneous Linear Equations Solver let you manually enter the syntax or let the wizard guide you through the process.

Figure 5-13:
The Algebra submenu commands.

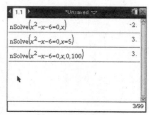

Numerical Solve

Polynomial Root Finder

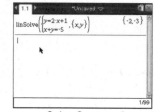

Solve System of Linear Equations

Probability and statistics

Press [menu]⇨Probability to access the Probability submenu.

Look at the first screen in Figure 5-14; notice I used the Factorial, Permutations, and Combinations commands in the first three lines. These three functions are invoked by first pressing [menu]⇨Probability. The last three lines in this screen are part of the Random submenu (press [menu]⇨Probability⇨Random).

Press [menu]⇨Statistics to access the Statistics submenu. The second screen in Figure 5-14 shows the Statistics submenu. Notice that all but the Stat Results choice have a third-level submenu.

Many of the items in this menu are used in conjunction with other applications, mainly the Lists & Spreadsheet application. Data is usually located in the Lists & Spreadsheet application. Additionally, some commands can be nested within other commands. The third screen in Figure 5-14 shows two such examples. The first line in this screen shows how to find the mean of 50 random integers from 1 to 6. The second line in this screen shows how to find the sum $1^2 + 2^2 + 3^2 + \ldots + 10^2$.

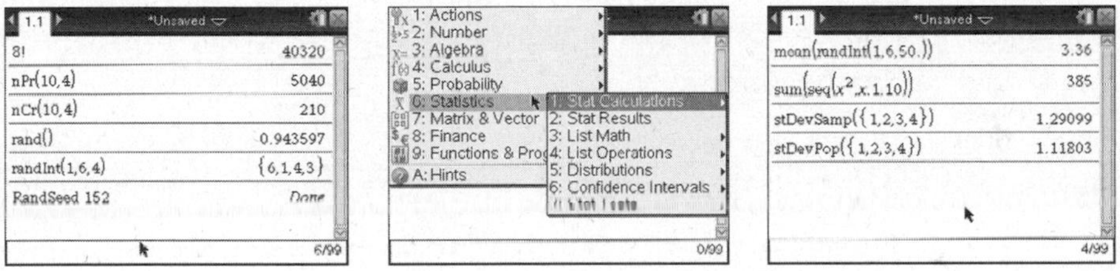

Figure 5-14: Probability submenu commands and the Statistics submenu.

Exploring matrices

Press menu⇨Matrix & Vector to access the Matrix commands.

Many of the items contained in the Matrix & Vector menu work with a matrix that you must first define. For example, check out the first screen in Figure 5-15. I pressed menu⇨Matrix & Vector⇨Determinant to paste the Det command to the entry line. The easiest way to define a matrix as the argument for the Det command is via the Expression templates. Follow these steps to accomplish this task:

1. **Press** [x] **to open the Expression templates.**

2. **Highlight the Variable Size Matrix template (see the first screen in Figure 5-15).**

3. **Press** enter **to open the Create a Matrix dialog box, as shown in the second screen in Figure 5-15. Use the ▲▼ keys to set the number of rows and columns.**

4. **Press** enter **to paste a blank matrix with the specified dimensions to the argument of the Det command.**

5. **Enter each element of the matrix, pressing** tab **each time to move to the next field.**

6. Press enter to execute the command.

See the last screen in Figure 5-15; I multiplied two matrices and used the Last Answer feature with the Transpose command to transpose the rows and columns of my answer.

Figure 5-15:
Creating a matrix and using the Matrix & Vector submenu.

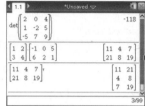

Working with the sophisticated Calculus and Finance tools

Even if you aren't in a calculus class, you may find some of the tools in the Calculus submenu helpful. Using the Numerical Function Minimum tool, you can find the minimum of a function without graphing it. The Sum template finds the sum of a sequence, and the Product template finds the product of a sequence. (See the first screen in Figure 5-16.)

To find the numerical derivative at a point, press menu⇨Calculus⇨Numerical Derivative at a Point. A wizard automatically starts to help you enter the syntax correctly. On the flip side, the Calculus submenu also has a Numerical Integral command, which uses a template to find the definite integral. (See the second screen in Figure 5-16.)

The Finance Solver is one of the most versatile tools on the TI-Nspire. In the third screen in Figure 5-16, I used the Finance Solver to calculate how many years it'd take to save a million dollars at 10 percent interest if I saved $5 a day (a little over 42 years). In addition, you can create an amortization schedule for a mortgage. And, using the most obscure tool on the TI-Nspire, you can find out how many days are left in the school year with the Days between Dates command.

Figure 5-16:
Calculus commands and the Finance solver.

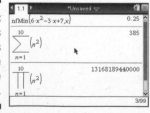
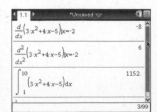
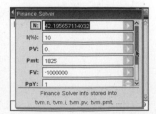

Chapter 6

Working with Variables

*T*his chapter is all about variables — how to create variables, name variables, and use variables to enhance your experience working with TI-Nspire. Keep in mind that variables are shared among applications within the same problem. The ability to share variables within a problem allows you to dynamically represent math concepts in multiple ways.

The Rules for Naming Variables

TI-Nspire offers a great deal of flexibility when naming variables. However, a few variable names are off-limits. Here's the official list of do's and don'ts when naming variables:

✔ Variables can be 1–16 characters long and may consist of letters, digits, and the underscore character.

✔ TI-Nspire does not distinguish between uppercase and lowercase letters.

✔ The first character cannot be a digit.

✔ Spaces are not allowed.

✔ If an underscore is used as the first character, the variable is considered a type of unit. Units do not allow subsequent underscores in the name.

✔ System variables, functions, and command names cannot be used as variables. Examples include Ans, fMax, and Mean.

Storing Variables

You might want to store a variable for a variety of reasons. Here's a list of some of those reasons:

- ✓ **Store a number:** You may want to store a number to a variable if you expect that you'll need to use the number in subsequent calculations. Storing a number is especially helpful if the number is irrational (or quite long) and you want to store the entire result for future use.

- ✓ **Define a function:** The ability to define a function with a meaningful name is a powerful option. In fact, TI-Nspire has dispelled the myth that functions must have names like $f(x)$ or $g(x)$. Now you can create functions whose names actually tell you what the function does. My two examples from this chapter — *area(s)* and *surface_area(r,h)* — illustrate this point.

- ✓ **Store a list:** Lists can be quite long and cumbersome. By storing a list as a variable, you can recall the list using a single variable name, rather than by retyping all the elements contained in the list.

- ✓ **Store a matrix:** Matrices can also be long and cumbersome, particularly if the matrix contains several rows and columns. Storing a matrix can save you a lot of time and effort.

In general, storing variables can be a great time-saver; it also ensures that subsequent actions that utilize stored variables are precise and accurate. Finally, there are many, many situations in which a stored variable can facilitate and enrich a mathematical exploration. You certainly see these types of applications as you read other chapters in this book.

Some variables are stored automatically depending on the functions you access while using TI-Nspire. For example, each time you perform a calculation, the last answer is stored to the variable Ans. If you perform a regression on a dataset, several variables are created, such as Stat1.r (correlation coefficient), stat.resid (list of residuals), and stat1.regeqn (regression equation). These examples also serve to remind you of the types of variables that TI-Nspire can store. The variable Stat1.r is a *number*. The other two variables, stat1.resid and stat1.regeqn, are a *list* and a *function,* respectively.

Variables can only be shared among pages that are part of the same problem. If you define a variable in Problem 1, this variable can be accessed only from within Problem 1. Furthermore, you can define a variable with the same name in a second problem, knowing that these two variables won't conflict with one another. They can take on completely separate values or meanings. Chapter 7 discusses the use of variables in applications other than the Calculator application. You can't create a global variable across all problems in the TI-Nspire.

REMEMBER

> A variable that is stored in a document can't be accessed from the Scratchpad (because the Scratchpad is not part of a document to begin with).

You also can decide what variables you want to store. The next two sections explain how.

Using the store variable ([ctrl] [var]) key

The Store Variable operator is a secondary key, accessed by pressing [ctrl] [var]. When using this method to store a variable, follow these three steps:

1. **Type the item, such as a value, list, matrix, or expression, you want to store on the entry line.**

2. **Press [ctrl] [var] to open the Store command (as indicated by a small right arrow).**

3. **Type the variable name and then press [enter] to store the variable.**

An alternate method for storing a variable

A colon followed by an equal sign ([ctrl] [=]) also tells TI-Nspire to store a variable. This method works just the opposite of the [ctrl] [var] method. That is, you must first type the variable name; press [ctrl] [=]; type the value, list, matrix, or expression to be stored; and then press [enter].

The first screen in Figure 6-1 shows several examples of how variables can be stored using the first method. The second screen in Figure 6-1 shows how these same variables can be stored using the second method.

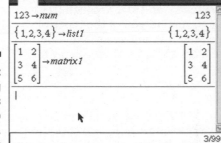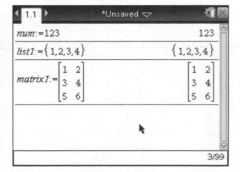

Figure 6-1: Storing variables using two methods.

Using the Define command

The Define command offers yet another way to store a variable. Here's an example of how you can create your own function in the Calculator application using the Define command:

1. **Press [menu]⇨Actions⇨Define to paste the Define command to the entry line.**

2. **Type the name of your function using function notation.**

   You must include the independent variable within parentheses.

3. **Press [=], type the function rule in terms of the independent variable that you've selected, and press [enter].**

The first screen in Figure 6-2 shows an example of a user-defined function that gives the area of an equilateral triangle, *area(s)*, with side length *s*. The second screen in Figure 6-2 shows an example of a user-defined function that gives the surface area of a cylinder, *surface_area(r,h)*, in terms of the input variables radius *r* and height *h*.

Keep in mind that these functions can also be defined using the methods described in the previous two sections. For example, you can define the function *surface_area(r,h)* by typing **surface_area(r,h):=2·Π·r²+2·Π·r·h**.

The third screen in Figure 6-2 shows how to evaluate these functions for specific values of the independent variables. You can access these user-defined functions by pressing the [var] key and selecting the defined function from the list. You can also type the function using the alphanumeric keys. I talk more about recalling a stored variable in the next section.

Figure 6-2:
Using the
Define
command
to store a
function.

Stored functions can be graphed in the Graphs application. However, you must specify *x* as the independent variable for this to work. For example, to graph *A(s)*, type **A(x)** next to the first available function in the Graphs application.

You can also use the Define command to store a variable as a number, list, or matrix. Just invoke the Define command, type the variable name, press ⊟, and type the number, list, or matrix.

Multiple variables can be stored simultaneously by separating each store command with a colon (press 🔲 and choose the colon). For example, num1:=10:num2:-20 stores 10 to the variable num1 and 20 to the variable num2.

Recalling a Variable

After a variable has been stored, you will inevitably want to recall the variable for use in a command, expression, or other application. In this section, I give you two ways to recall a stored variable.

Typing a variable name

If you know the name of the variable you want to use, feel free to type the variable name using the alphanumeric keys. This method generally works fine if the variable name is relatively short and doesn't contain fancy symbols that are hard to find.

As shown in the first two screens in Figure 6-3, I used this method to recall the num variable. As you type, the letters appear in italics. When the variable name is complete, the entire variable name turns nonitalicized and bold (see first screen in Figure 6-3). Press ⏎ to see the value of the variable as shown in the second screen in Figure 6-3.

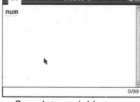
Complete variable turns bold, non-italic

The value of the variable

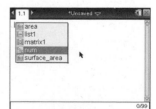
Access the variable from a list

Using the [var] *key*

The [var] key provides a quick way to access variable names. Press [var], scroll through the variable list using the Touchpad keys to highlight your choice, and press [enter] to paste it to your current location. The third screen in Figure 6-3 shows an example of the list that appears after you press the [var] key.

When pressing the [var] key, small icons located to the left of each variable name indicate the variable type. Table 6-1 gives the meaning of each icon.

Table 6-1	Stored Variable Icons
Icon	*Description*
ꝼꞏꞏ	Function
ꞁꞏꞏꞁ	List
ꞁꞏꞏꞁ	Matrix
⁰ꞁ₂	Number

The two methods described in this section for recalling variables work in any TI-Nspire application.

Updating a variable

Say you just stored a variable with a certain value (or function, matrix, or list). If you store the variable again with a new value (or function, matrix, or list), the variable takes on this new value.

You can also use the variable itself to define the same variable with a new value.

The first screen in Figure 6-4 shows some examples in which variables are stored to new values. In the first line, 10 is stored to the variable num. On the second line, 16 is stored to the same variable. On the third line, num is recalled to show that it's taken on this new value. On the fourth line, a more complicated expression contains num and stores the result, 248, back to num. On the last line, num is recalled to show that it's taken on this new value.

The second screen in Figure 6-4 shows some examples of how a stored matrix can be updated using the alternative method of storing variables.

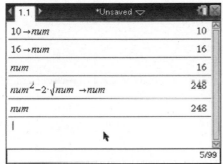

Figure 6-4:
Updating
variables.

Deleting Variables

Variables can be deleted from within the Calculator application via the DelVar command (press menu⇨Actions⇨Delete Variable). Keep in mind that you can delete a variable that's contained only within the current problem. The DelVar command can be used to delete several variables at a time by separating each variable to be deleted with a comma.

Figure 6-5 shows a few ways to delete variables using the DelVar command.

Figure 6-5:
Deleting
variables.

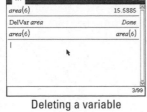  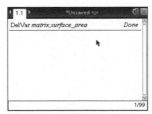

Deleting a variable   Pressing the variable key   Deleting multiple variables

After you type the DelVar command, press ⟨var⟩ to bring up a list of stored variables and then select the variable to be deleted from this list. If you call up a function, it also brings up a set of parentheses, which must be deleted by pressing ⟨del⟩ once.

Don't delete a variable unless you really mean to! It's one of the few times that the Undo command doesn't work.

Chapter 7

Using the Calculator Application with Other Applications

....

In This Chapter

▶ Accessing variables from within any application

▶ Connecting between the Calculator app and the Graphs and Geometry apps

▶ Using the Calculator app with the Lists & Spreadsheet app

....

*Y*ou can do so much by working strictly within the Calculator application. However, this application doesn't work in isolation. In this chapter, I show you some examples that illustrate how the Calculator application can work in conjunction with other applications.

Defining Variables to Be Used Elsewhere

If you define a variable, it's available for use in any page in TI-Nspire, assuming that you work on pages within the same problem. To recall the variable, you have two options:

▶ **Press var and select the desired variable from the list.**

▶ **Type the variable name using the keypad.** If you choose this option, the variable name displays as nonitalicized and bold after the last character is typed, which indicates that you're working with a stored variable.

Using the Calculator Application with the Graphs Application

The Calculator application and the Graphs application are a perfect fit. In this section, I show you several examples of how you can establish the lines of communication between these two powerful applications.

Defining a function and graphing it in the Graphs application

In the following list, I show you how to define a function in the Calculator application and then how to graph it in the Graphs application.

1. **Use the Define command to define *area(r)*.**

   In the first screen in Figure 7-1, I used the circle area formula.

2. **Press `ctrl` `I` and open a new Graphs page.**

   By default, you're in graphing mode with the cursor located on the entry line next to the first available function.

3. **Press `var` to view a list of variables and select *area* from the list. (See the second screen in Figure 7-1.)**

   You can also type the variable name using the alpha keys. If you do, don't forget to place an open parenthesis after the variable.

4. **Type the variable *x* as the independent variable (even though you used *r* when defining the function).**

5. **Press `enter` to graph the function and adjust the window settings accordingly.**

   I added some lines and included a moveable point, with coordinates, on the graph. (See the third screen in Figure 7-1.)

Figure 7-1: Graphing a function defined in the Calculator application.

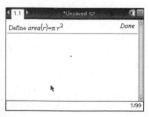  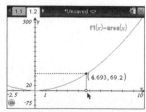

TIP

Because you graphed the function *area(r)* in the Graphs application as *f1(x)*, you can now access this function in two ways, *area* and *f1*. For example, from within a Calculator page, you can type **area(4)** to find the area of a circle with radius 4. Alternatively, you can evaluate f1(4) and obtain the same result.

Graphing a function with two input variables

Now, for a much tougher challenge: defining a function, *volume(r,h)*, that gives the volume of a cone as a function of the radius, *r*, and height, *h*. See the first screen in Figure 7-2. This function is trickier to graph because there are two input variables. To get around this issue, create a slider in the Graphs application to give the height, *h*, a fixed value with the option of changing this height by dragging the slider. In effect, you're defining *h* as a constant but still giving yourself the option of varying it. See Chapter 9 to find out how to use a slider.

The second screen in Figure 7-2 shows the graph of this scenario with *h* set to 2.5. The third screen in Figure 7-2 shows this same scenario with *h* changed to 4.5. Notice that the point on the graph corresponding to *x* = 7 has corresponding *y*-values of 128.28 and 230.91 for *h* = 2.5 and *h* = 4.5, respectively.

Figure 7-2:
Graphing
a function
with two
independent
variables.

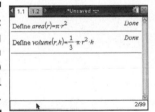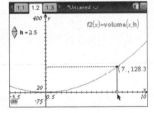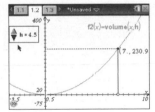

Evaluating a function in the Graphs application

To evaluate a function defined in the Graphs application at a specific value for the input variable, follow these steps:

1. **Type the name of the function using the keypad or press [var] to call up a list of defined functions.**

2. **After you type the function name, press [(] to open a set of parentheses.**

   If you access a function using the [var] key, a set of parentheses is included automatically.

3. **Type any numerical value within the parentheses and press [enter] to evaluate the function for this value.**

   If you mistakenly attempt to evaluate a function for a value for which it's not defined, an error message appears.

Graphing a step function in the Graphs application

Many other functions can be graphed by accessing commands commonly used in the Calculator application. The *greatest integer function,* $y = \text{int}(x)$, is one such graph, which you create by following these steps:

1. **Position the cursor next to the first available function line.**

2. **Press [⊞] to open the Catalog.**

   If necessary, press [1] to activate the first category containing the alphabetical listing of all commands, functions, and symbols.

3. **Press [1] to jump to the items beginning with I, scroll down, highlight int, and press [enter] to paste this command to the Graphs entry line.**

4. **Type x for the argument and press [enter] to complete the graph.**

The first screen in Figure 7-3 shows the graph of the greatest integer function, and the second screen shows the graph of the absolute value function (which uses the Abs command from the Catalog). Alternatively, you could use the Absolute Value Expression template to type the function.

Figure 7-3:
Using other
functions
from the
Calculator
application
for
graphing.

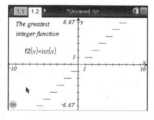 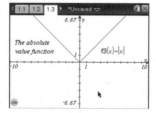 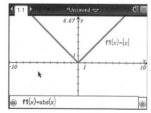

Rather than use the Catalog, you can type a function directly using the alpha keys. That's how I graphed the absolute value function in the third screen in Figure 7-3. I simply typed [A][B][S], and the command was recognized as indicated by the nonitalicized font. After pressing [(][X][enter], the Abs command is replaced by the vertical bars that are used to denote absolute value. This is a handy trick when using TI-Nspire computer software.

Using the Calculator Application with the Lists & Spreadsheet Application

In the earlier sections, I tell you how the Calculator application and the Graphs and Geometry applications talk to one another. It turns out that the Calculator application is very social, talking to any application, including the Lists & Spreadsheet application.

Storing lists from the Calculator application to L&S

The Calculator app contains a variety of commands that are useful in generating lists of data. One such example is the Sequence command, accessed by pressing [menu]⇨Statistics⇨List Operations⇨Sequence. The syntax for this command is seq(*Expression, Variable, Low, High*[, *Step*]. For example, the command seq(2x + 1,x,0,50,5) generates the list {1, 11, 21, 31, ..., 101}. The Random submenu (press [menu]⇨Probability⇨Random) offers another place where lists of data can be produced.

If you're interested in investigating the outcomes of rolling two dice 50 times, here's how to accomplish this task:

1. **Press [ctrl][I], select a Lists & Spreadsheet page and name the first column *red_die* and the second column *blue_die*.**

2. **Press [ctrl][I], select a Calculator page and press [menu]⇨Probability⇨ Random⇨Integer.**

   This pastes the command randInt to the entry line.

3. **Configure this command to read randInt(1,6,50) and then press [ctrl][var] to open the Store command.**

4. **Press [var], highlight *red_die* from the list, and press [enter].**

   See the first screen in Figure 7-4.

5. **Press [enter] again to execute the command.**

   You see the list of numbers displayed on the Calculator page. (See the second screen in Figure 7-4.) This data is also stored to the Lists & Spreadsheet page under the *red_die* column.

6. **Repeat Steps 2–5 and store 50 random integers from 1 to 6 to the *blue_die* list.**

   See the third screen in Figure 7-4.

Figure 7-4:
Storing two-dice data to the Lists & Spreadsheet application.

Storing random integers to *red_die*

Results displayed on the Calculator page

Results stored to the Lists & Spreadsheet application

Performing statistical analyses on data

Data contained within the Lists & Spreadsheet application can be analyzed from within the Calculator application. For example, you might be interested in viewing the one-variable statistical results of the Total list from the two-dice sum experiment. Follow these steps:

1. **Press [menu]⇨Statistics⇨Stat Calculations⇨One-Variable Statistics.**

2. **Press [enter] to indicate that you want to analyze one list.**

   If you have additional lists, change Num of Lists to match the number of lists you're interested in analyzing.

3. **Configure the dialog box by choosing X1 list total, as shown in the second screen in Figure 7-5.**

4. **Press [enter] to close the dialog box and perform the statistical analysis. (See the third screen in Figure 7-5.)**

Figure 7-5:
Performing a one-variable statistical analysis from within the Calculator app.

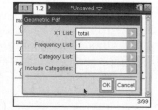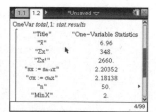

Other variations on the two-dice example

In the nearby "Storing lists from the Calculator application to L&S" section, I show you how to execute two separate commands to store 50 random integers from 1 to 6 to two separate lists. You can execute both of these commands on a single entry line by separating them with a colon. The exact syntax for this command is `randint(1,6,50)→red_die:randInt(1,6,50)→blue_die`. Press `enter` to populate both lists in the Lists & Spreadsheet page with 50 random integers from 1 to 6 simultaneously.

To find the sum of these die, name the third column in the spreadsheet *Total*. Then go back

to the Calculator page, type the command **red_die + blue_die→total**, and press `enter`. Take a look at the Lists & Spreadsheets page. The third column contains 50 numbers, with each value equal to the sum of the two numbers found in the first and second columns. Refer to the first screen in Figure 7-5.

If you want to generate an entirely new set of two-dice data, use the Calculator history to re-execute the `randInt(1,6,50)→red_die:randInt(1,6,50)→blue_die` and `red_die + blue_die→total` commands. See Chapter 5 for more information on how to use the Calculator history.

A variety of statistical results are generated by the One-Variable Statistics command. See Chapter 14 for a description of what each result means.

Many of the statistical results generated by the One-Variable Statistics command can be opened individually from within the List Math submenu (press `menu`⇨Statistics⇨List Math). For example, the command Sum of Elements (denoted Sum) yields the sum of all the elements of a specified list. In general, all the commands found within the List Math submenu must contain either a list or a list name as their arguments.

You can sort lists or perform a variety of other manipulations of lists from within the Calculator application. Many of the commands that allow you to perform such manipulations are found in the Lists Operations submenu (press `menu`⇨Statistics⇨List Operations). For example, the command Sort Ascending (denoted SortA) sorts a specified list from lowest value to highest value.

Performing regressions

In Chapter 15, I discuss how to perform a regression from within the Lists & Spreadsheet application. In Chapter 19, I show how to perform a regression from within a Data & Statistics page (the time-saving choice). You can also perform a regression on data contained in the Lists & Spreadsheet application from within the Calculator application.

In the first screen of Figure 7-6, I used a Lists & Spreadsheet page to put five data points into x_list and y_list lists. (To find out more about how to configure the Lists & Spreadsheet application for this data, see Chapter 15.) This figure also shows a Quick Graph of the data, which suggests that an exponential model is a good choice.

To perform an exponential regression on this data in the Calculator application, follow these steps:

1. **Press** menu⇨**Statistics**⇨**Stat Calculations**⇨**Exponential Regression.**

   A dialog box opens, as shown in the second screen in Figure 7-6. As with any dialog box, you can press tab to move from one field to the next or ⇧shift tab to move backward through a field.

   Configure the dialog box by choosing x_list for the X list and y_list for the Y list. You can type these names using the alpha keys or press and select x_list to specify the location of the X List and select y_list for the Y List.

2. **Press** enter **after configuring the dialog box to perform the regression.**

   All the results of the regression paste to the Calculator page, as shown in the third screen of Figure 7-6.

Figure 7-6:
Performing
an expo-
nential
regression
from within
the Calculator
application.

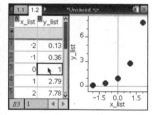
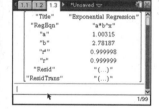

The exponential regression results yield the equation $f2(x) = (1.00315)(2.78187)^x$. Notice the *Coefficient of Determination*, R^2, is very close to 1. This might suggest that the regression equation is a good fit for the data.

A variety of variables are stored by TI-Nspire after a regression. To view this list, press the var key. Using the ▲▼ keys, you can scroll through this list and paste a variable to the entry line in the Calculator page.

For example, try selecting the variable Stat.Resid and pressing enter to paste it to the entry line. Press enter again to produce a list of *residuals.* This list represents the difference between the *y*-value of each data point and the corresponding *y*-value associated with the regression equation.

Chapter 8

Using the Calculator Application with TI-Nspire CAS

The acronym *CAS* stands for *computer algebra system*. A computer algebra system facilitates the symbolic manipulation of mathematical expressions and equations. For example, consider the solution to the equation $x^2 = 12$. A computer algebra system, such as the one built into the TI-Nspire CAS Handheld, returns the answers

$$x = -2\sqrt{3}$$

and

$$x = 2\sqrt{3}$$

The TI-Nspire numeric performs numerical or *floating-point* calculations. Hence, the solutions to the equation $x^2 = 12$ are given as –3.464 and 3.464.

Throughout this book, I refer to the TI-Nspire numeric handheld as simply TI-Nspire. I refer to the CAS handheld as TI-Nspire CAS.

In this chapter, I demonstrate the symbolic manipulation capabilities of TI-Nspire CAS as they apply to the Calculator application. In Chapter 10, I discuss how TI-Nspire CAS can be used with the Graphs application.

TI-Nspire and TI-Nspire CAS have a significant amount in common, and just about everything you read in this book most certainly applies to both.

A brief history of computer algebra systems

Computer algebra systems were first invented in the early 1970s. Examples of computer-based CASs include Maple, Mathematica, Derive, and MathCAD. Derive requires relatively little memory and processing power. As a result, it can be used on older machines with limited capabilities. Furthermore, Derive helped make possible the introduction of the TI-92 in 1995, one of the first devices that offered computer algebra capabilities in handheld form. Texas Instruments introduced the successor to the TI-92, the TI-89 series, in 1998. The TI-89 is similar to the TI-92 except that it is smaller and does not have a QWERTY keyboard. As a result, the TI-89 is allowed on most standardized tests. Texas Instruments introduced TI-Nspire CAS in 2007. TI-Nspire and TI-Nspire CAS are also allowed on most standardized tests. See www.education.ti.com for the latest on assessment policies and calculator usage on tests.

Evaluating Expressions

In this section, I show you how the results of simple calculations are given, by default, symbolically. I then show you how to force approximate results.

Finding symbolic representations of numerical calculations

Check out the first screen in Figure 8-1. I've entered some expressions by using a combination of primary and secondary keys. As you can see in the first two lines of these screens, TI-Nspire CAS returns results in symbolic form. That is, results are given as exact values — the way you typically see them in textbooks or other printed materials. On the third line of this same screen, you can see that TI-Nspire CAS attempts to display algebraic expressions in simplified form. In the case of the product of two rational expressions, common factors are divided out, and the result is displayed as a single, simplified rational expression. TI-Nspire CAS can even handle complicated expressions such as the one shown in the last line of the first screen in Figure 8-1. Notice that warning message located at the bottom of the first screen in Figure 8-1. The complete message reads "Domain of the result might be larger than the domain of the input." This message occurs because the domain of the input is the set of all real numbers greater than zero, whereas the domain of the output is the set of all real numbers greater than or equal to zero.

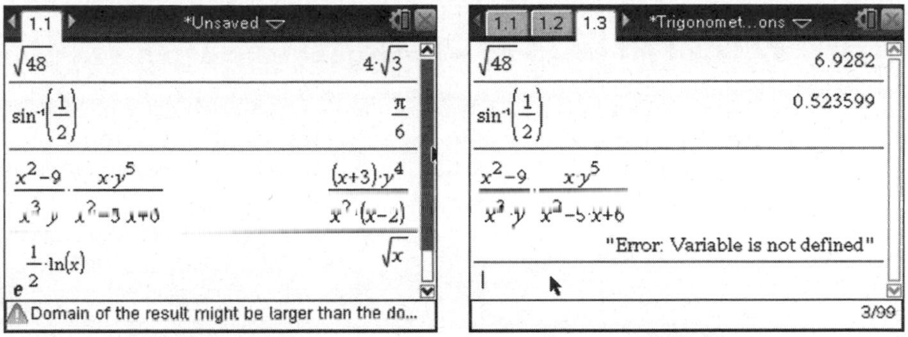

Figure 8-1: Comparing evaluated expressions on TI-Nspire CAS and TI-Nspire.

1.1 ▶ *Unsaved ▽	
$\sqrt{48}$	$4\cdot\sqrt{3}$
$\sin^{-1}\left(\dfrac{1}{2}\right)$	$\dfrac{\pi}{6}$
$\dfrac{x^2-9}{x^3\cdot y}\cdot\dfrac{x\cdot y^5}{x^2-3\cdot x+9}$	$\dfrac{(x+3)\cdot y^4}{x^?\cdot(x-2)}$
$\dfrac{1}{e^{\frac{1}{2}\cdot\ln(x)}}$	$\sqrt{x}$

⚠ Domain of the result might be larger than the do...

TI-Nspire CAS

1.1 1.2 1.3 ▶ *Trigonomet...ons ▽	
$\sqrt{48}$	6.9282
$\sin^{-1}\left(\dfrac{1}{2}\right)$	0.523599
$\dfrac{x^2-9}{x^3\cdot y}\cdot\dfrac{x\cdot y^5}{x^2-5\cdot x+6}$	
	"Error: Variable is not defined"
	3/99

TI-Nspire

As a comparison, I've typed the first three expressions on the TI-Nspire Handheld, as shown in the second screen in Figure 8-1. The first two results are given as decimal approximations. The rational expression on the third line returns an error message because the variables x and y are not defined and, therefore, do not have a numerical value. Had I stored numerical values to x and y, TI-Nspire would return the numerical result of this expression evaluated at these stored values.

TI-Nspire and TI-Nspire CAS do not always display results as decimals. Answers that are rational numbers are almost always given in fractional form. However, the examples shown in the second screen in Figure 8-1 have irrational answers and, therefore, must be given as decimal approximations.

Finding approximate results

At times, it is advantageous to view the decimal approximation of an answer. As you do with TI-Nspire, press [ctrl] [enter] to evaluate an expression and force a result as a decimal. Here are three additional ways to obtain an approximate result:

- ✔ **Include a decimal point somewhere in your calculation.** For example, press [3][+][3][÷][5][.] to evaluate the first example in decimal form.

- ✔ **Use the Convert to Decimal command.** Press [3][+][3][÷][5] and then press [menu]⇨Number⇨Convert to Decimal. This command can also be found in the Catalog.

- ✔ **Use the Approx command to force a result in decimal form or scientific notation.** This command can be typed using the alpha keys and can also be found in the Catalog.

The TI-Nspire CAS Application Menu

Take a look at the screens shown in Figure 8-2. These screens represent the Algebra submenus for both TI-Nspire devices. The TI-Nspire CAS has all the commands that TI-Nspire has, plus some additional commands. In this chapter, I focus mostly on these additional commands.

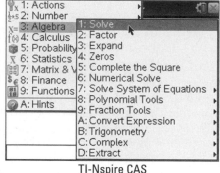
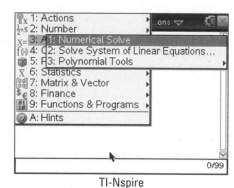

Figure 8-2: The TI-Nspire CAS and TI-Nspire Algebra submenus.

TI-Nspire CAS   TI-Nspire

Doing algebra

Earlier in this chapter, I demonstrated how TI-Nspire CAS can perform symbolic manipulation of numerical and algebraic expressions. As impressive as that is, the CAS capabilities found in the Algebra submenu are nothing short of amazing. In the following sections, I highlight some of the functions and commands in the Algebra submenu. Keep in mind that the name associated with each function reveals much about what it can accomplish.

The Solve command

Choose [menu]⇨Algebra⇨Solve to open the Solve command. As the name implies, this command returns the solution(s) to an equation or inequality.

Figure 8-3 shows some examples of what this command can do.

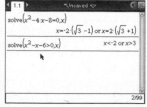
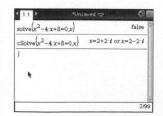
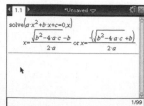

Figure 8-3: Using the Solve and cSolve commands.

The first screen in Figure 8-3 shows how the Solve command can solve a single equation or an inequality, with solutions displayed in symbolic form.

TIP

The first line in the second screen in Figure 8-3 shows the result of using the Solve command on equations with no real number solutions. The equation $x^2 - 4x + 8 = 0$ has no real number solutions. However, it does have two *complex* solutions. To find complex solutions, press menu⇨Algebra⇨Complex⇨Solve. This command, denoted cSolve, returns the complex solutions $x = 2 + 2i$ and $x = 2 - 2i$ for this equation.

TIP

Try solving the equation $a \cdot x^2 + b \cdot x + c = 0$ for the variable x by typing **solve($a \cdot x^2 + b \cdot x + c = 0, x$)**. TI-Nspire CAS returns the quadratic formula! Make sure that you press ⊠ between variables. Otherwise, TI-Nspire CAS may mistakenly think that ax and bx are single variables. See the third screen in Figure 8-3.

The Factor command

Choose menu⇨Algebra⇨Factor to open the Factor command. This command factors numerical and algebraic expressions. Take a look at the first two lines of the first screen in Figure 8-4. TI-Nspire CAS attempts to factor any expression as much as possible with linear, rational, and real factors. The expression shown in the second line is not factorable based on these conditions.

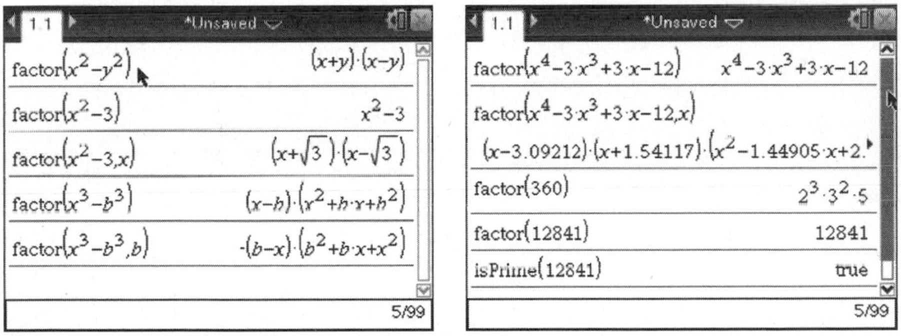

Figure 8-4: Using the Factor command.

◀ 1.1 ▶	*Unsaved ▽
factor(x^2-y^2)	$(x+y) \cdot (x-y)$
factor(x^2-3)	x^2-3
factor(x^2-3,x)	$(x+\sqrt{3}) \cdot (x-\sqrt{3})$
factor(x^3-b^3)	$(x-b) \cdot (x^2+b \cdot x+b^2)$
factor(x^3-b^3,b)	$-(b-x) \cdot (b^2+b \cdot x+x^2)$
	5/99

◀ 1.1 ▶	*Unsaved ▽
factor$(x^4-3 \cdot x^3+3 \cdot x-12)$	$x^4-3 \cdot x^3+3 \cdot x-12$
factor$(x^4-3 \cdot x^3+3 \cdot x-12,x)$	
$(x-3.09212) \cdot (x+1.54117) \cdot (x^2-1.44905 \cdot x+2.$ ▸	
factor(360)	$2^3 \cdot 3^2 \cdot 5$
factor(12841)	12841
isPrime(12841)	true
	5/99

Notice in the third and fourth lines that I've included a comma followed by a variable. By specifying a variable, TI-Nspire CAS produces linear factors (that is, factors in which the degree of the specified variable is 1) as long as these factors contain only real numbers.

The fourth and fifth lines are intended to illustrate that TI-Nspire CAS provides some sorting when it does its factoring. For example, notice that the fourth line is sorted with respect to the variable x (by *sorted*, I mean that x shows up first in the factors). In the fifth line, I've specified that I want to factor with respect to the variable b. As a result, the factors in the result lead with this variable.

Now, look at the second screen in Figure 8-4. The first line returns the original expression because the polynomial $x^4 - 3x^3 + 3x - 12$ is *prime*, meaning that it cannot be factored into linear, rational factors. However, by specifying that I want to factor in terms of the variable x (as shown in the second line), I make TI-Nspire CAS produce linear factors with decimal approximations of irrational numbers, which provides a nice way to find the zeroes of a polynomial.

Finally, take a look at the third line of this screen, which illustrates that the Factor command factors numbers, too. The fourth line gives an example of what you get when you try to factor a prime number.

TIP

The computing time required to factor large composite numbers can be quite long. If you are interested only in determining whether a number is prime or composite, try using the isPrime command. This command returns the word *true* if the number is prime and *false* if the number is composite. See the fifth line of the second screen in Figure 8-4.

The Expand command

Choose [menu]⇨Algebra⇨Expand to open the Expand command. This command works just the opposite of the Factor command. It multiplies out expressions in factored form, including those containing exponents.

As you can see in the first screen in Figure 8-5, other similarities exist between the Expand and Factor commands. If you specify a variable to expand with respect to, TI-Nspire CAS adjusts the order in which the variables are presented.

Figure 8-5:
Using the
Expand
command.

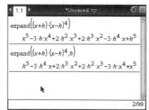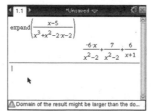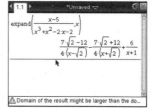

The second screen in Figure 8-5 illustrates that the Expand command also gives the partial fraction expansion for rational expressions. In the third screen, I've used the Expand command again with the same rational expression. However, I've also specified the variable x. Because I did so, TI-Nspire CAS provides a more complete partial fraction expansion.

Other commands found on the Algebra submenu

The Solve, Factor, and Expand commands form the cornerstone of the Algebra submenu. However, several other items contained on the Algebra submenu are worth mentioning.

In the first screen in Figure 8-6, I have accessed the Solve a System of Equations tool, which automatically starts a wizard. I recommend using this tool to solve systems of equations. Technically, the wizard correctly places the syntax for the Solve command. I could have accessed the Solve command directly, but why try to remember the syntax if I can use a wizard? The second line of the second screen shows how the solution to a *dependent* system is displayed. The infinite number of solutions is indicated by an expression in terms of the constant $c2$. This constant can equal any real number and gives the set of ordered pairs that lie on the line $2x - y = -3$.

Figure 8-6: A few highlights from the Algebra submenu.

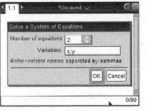 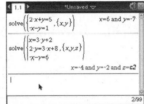

The third screen showcases some of my favorite tools. To open the Proper Fraction tool I pressed [menu]⇨Algebra⇨Fraction Tools⇨Proper Fraction. This tool is very useful for doing long division of polynomials or finding the slant asymptote of a rational function. I also show another fraction tool (only available on a TI-Nspire CAS), the Common Denominator tool. The newest tool in the Algebra submenu is the Complete the Square tool which puts a quadratic equation into vertex form for a parabola.

The following is a list of some of the other functions found on the Algebra submenu and a brief description of what they do:

TIP

✔ **Zeros:** The syntax for this command is `Zeros(Expression, Variable)`. This command produces a list of the values of the specified *variable* that make the *expression* equal zero.

Sometimes the Zero command returns the empty set because the zeros are complex. To find complex zeros, press [menu]⇨Algebra⇨Complex⇨Zeros.

✔ **Polynomial tools:** This submenu contains the following polynomial commands:

- *Remainder of Polynomial:* The syntax for this command is `polyRemainder(Expression1, Expression2)` and returns the remainder when *Expression1* is divided by *Expression2*.

- *Quotient of Polynomial:* The syntax for this command is `polyQuotient(Expression1, Expression2)` and returns the quotient when *Expression1* is divided by *Expression2*, less the remainder.

- *Greatest Common Divisor:* The syntax for this command is `polyGcd(Expression1, Expression2)` and returns the greatest common rational factor of *Expression1* and *Expression2*.

- *Coefficients of Polynomial:* The syntax for this command is `polyCoeffs(Expression)` and returns the numerical coefficients, in list form, of the polynomial *Expression*. Coefficients are given in order from the highest degree term down to the lowest degree term. For example, `polyCoeffs(5x + x`^{3}` - 3)` returns the list {1, 0, 5, -3}. Notice that a zero is given for the missing x^2 term.

- *Degree of Polynomial:* The syntax for this command is `polyDegree(Expression)` and returns the degree of the polynomial.

✔ **Convert Expression tools:** This submenu contains the following commands:

- *Convert to ln:* The syntax for this command is `Expression` ¢**ln** and returns an expression only containing natural logs.

- *Convert to logbase:* The syntax for this command is `Expression1` ¢**logbase**(`Expression2`) and returns a simplified expression using base *Expression2*.

- *Convert to exp:* The syntax for this command is `Expression` ¢**exp** and returns an expression in terms of the natural exponential, *e*.

- *Convert to sin:* The syntax for this command is `Expression` ¢**sin** and returns an expression in terms of sine.

- *Convert to cos:* The syntax for this command is `Expression` ¢**cos** and returns an expression in terms of cosine.

✔ **Trigonometry:** This submenu contains the following trigonometry commands:

- *Expand:* The syntax for this command is tExpand(*Expression*) and returns the expansion of sines and cosines whose angles are integer multiples, sums, and differences. For example, the expression tExpand(sin(α+θ) returns the angle-sum identity cos(α)·sin(θ)+sin(α)·cos(θ). This is a great command to use if you forget the sum, difference, double, and half-angle trigonometric formulas.

- *Collect:* The syntax for this command is tCollect(*Expression*) and returns an expression in which powers and products of sines and cosines are converted to linear combinations of sines and cosines of multiple angles, angle sums, and angle differences. Basically, the tCollect command reverses the results obtained by the tExpand command. For example, the expression tCollect(2·(cos(θ))²−1 gives cos(2·θ) as a result.

The α symbol, as well as other Greek characters, can be found in the Symbol palette (ctrl ⌨). The θ character can also be found by pressing π· and choosing θ.

Exploring calculus using CAS

Press menu ⇨ Calculus to access the Calculus submenu, as shown in the first screen in Figure 8-7. Many of these functions can also be accessed via the Expression Template, as shown in the second screen in Figure 8-7. Starting with the first icon to the right of the shaded icon in the Expression templates, you have the Sum template, Product template, First Derivative template, Second Derivative template, Nth Derivative template, Definite Integral template, Indefinite Integral template, and Limit template.

Figure 8-7:
Accessing
Calculus
functions.

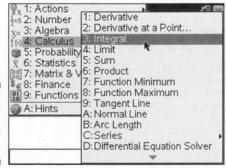
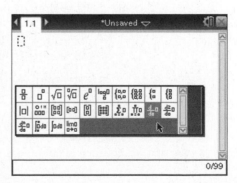

In the Calculus submenu, the CAS technology allows symbolic manipulation of algebraic expressions as well as numerical calculations just like in the Algebra submenu. You'll understand what I mean by this as you read this section.

The Derivative, Integral, and Limit commands form the cornerstone of the Calculus submenu, and I provide some detail as to how these functions work in the next three sections. I also provide a brief overview of some of the other items contained within the Calculus submenu.

Taking derivatives

Press menu⇨Calculus⇨Derivative to open the Derivative command. A template containing two fields is pasted to the entry line. The current active field (as indicated by the blinking cursor) allows you to type the variable that you are finding the derivative with respect to. Type this variable and then press tab to move to the second field enclosed in parentheses. Type the expression that you want to find the derivative of here and press enter to find the derivative.

Figure 8-8 provides some examples of how to use the Derivative command. Here are a few comments about the first screen in Figure 8-8:

- ✓ **The first line shows a common use of the Derivative command.** Notice that TI-Nspire CAS displays an answer identical to what might appear in a textbook.

- ✓ **In the second line, I typed Xx^2 in the first field.** This tells TI-Nspire CAS to give the second derivative. To find higher-order derivatives, press ^ followed by the derivative order.

- ✓ **The** third line gives an alternative method of finding higher-order derivatives. I simply nest a number of derivative commands equal to the derivative order.

Here are some comments about the second screen in Figure 8-8:

- ✓ **In the first line, I typed X^−1 in the first field.** This tricks TI-Nspire CAS into giving the anti-derivative of the expression contained in the second field.

- ✓ **In the second line, I show you that TI-Nspire CAS can use function notation to provide the symbolic representation of the product rule.**

- ✓ **In the third line, I show you how to find the derivative of a list of expressions.** Make sure that you separate each expression with a comma and enclose the entire list in curly braces.

Finally, here are some comments about the third screen in Figure 8-8:

- ✔ **The first line shows TI-Nspire CAS's attempt at finding the symbolic rule for the quotient rule.**

- ✔ **In the second line, I've nested the derivative command in the Common Denominator command to obtain the quotient rule in a form consistent with what is found in many textbooks.** This is another important reminder that you can mix and match commands as needed.

Figure 8-8:
Using the
Derivative
command.

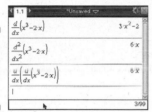
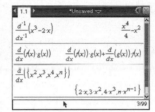
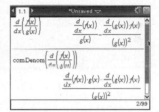

Working with integrals

Press menu⇨Calculus⇨Integral to open the Integral command. A template containing four fields is pasted to the entry line. The two leftmost fields are where you specify the lower and upper limits of integration. Leave these fields blank if you want to evaluate an indefinite integral. Type the expression to be integrated in the field contained in the parentheses. The rightmost field is where you type the variable to integrate with respect to.

Figure 8-9 provides some examples of how to use the Integral command. Here are a few comments about the first screen in Figure 8-9:

- ✔ **In the first line, the lower and upper limits of integration are omitted and the anti-derivative, less the constant of integration, is given.**

- ✔ **In the second line, I've included an expression for which the indefinite integral cannot be represented analytically.** As a result, the coefficient, *a*, is written before the integral symbol, and the expression containing *e* is left in integral form.

- ✔ **The third line shows that, although the indefinite integral cannot be found for the same expression containing the number *e*, a corresponding definite integral can be evaluated.**

Here are two comments about the second screen in Figure 8-9:

- ✔ **The first line shows that TI-Nspire CAS provides an exact answer for a definite integral whenever possible.**

- ✔ **The second line shows that I can force a decimal approximation by pressing** ctrl enter.

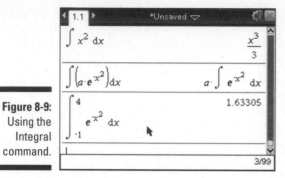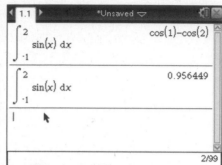

Figure 8-9:
Using the
Integral
command.

Evaluating limits

Press menu⇨Calculus⇨Limit to invoke the Limit command. A template containing four fields is pasted to the entry line. Type the variable in the leftmost field. Type the approaching value of the variable in the next field. Type ⊞ or ⊟ in the next field if you want to evaluate a one-sided limit. Leave this field blank if you want to evaluate the limit from both sides. Type the expression in the rightmost field.

Figure 8-10 provides some examples of how to use the Limit command. Here are a few comments about the first screen in Figure 8-10:

- **In the first line, I evaluate the limit of $x^2 - 3x + 1$ as x approaches 2.**

- **In the second line, I've used the definition of the derivative to find the derivative of the tangent function.**

- **In the third line, I've evaluated a limit at infinity.** Notice that TI-Nspire CAS does a great job of displaying the exact answer of this result. I accessed the symbol for the infinity character by pressing ctrl CA. You can also access the infinity symbol by pressing π· and choosing the infinity symbol.

Here are two comments about the second screen in Figure 8-10:

- **The first two lines show how to evaluate a left- and right-sided limit, respectively.**

- **The last line shows that the general limit of this absolute value expression is not defined.**

Figure 8-10:
Using the
Limit
command.

Other commands found on the Calculus submenu

Here are some other items contained on the Calculus submenu that are worth mentioning, including a brief description of what they do:

- **Tangent Line (& Normal Line):** The syntax for this command is `tangentLine(Expression, Variable, point)`. This command returns the equation of the tangent line to the equation $y = expression$ for an x-value equal to $point$. The Normal Line command works exactly the same way, except that it returns the equation of the line perpendicular to the tangent line at the same $point$ on the curve $y = expression$.

- **Arc Length:** The syntax for this command is `arclen(Expression, Variable, Start, End)`. This command gives the arc length of a function (given by $Expression$) from $x = Start$ to $x = End$. Try evaluating `arclen(f(x), x, a, b)` to obtain the textbook formula for the arc length along a curve.

- **Differential Equation Solver:** The syntax for this command is `deSolve(1st or 2nd Degree Differential Equation, Independent Variable, Dependent Variable)`. The command `deSolve(y¢=a×y,x,y)` returns the general solution $y = cke^{a \times x}$, where **k** is an integer suffix from 1 to 255. The first time you execute this command, TI-Nspire CAS returns the solution $y = c1e^{a \times x}$, where **c1** is an arbitrary constant. Subsequent occurrences of this arbitrary constant are denoted **c2**, **c3**, and so on. To find a particular solution, include the initial condition(s) with the differential equation. For example, the command `deSolve(y¢=a×y and y(0)=1,x,y)` returns the solution $y = e^{a \times x}$.

> Press ⌨ and choose the apostrophe to denote a first derivative. Press ⌨ and choose the apostrophe again to denote a second derivative.

- **Implicit Differentiation:** This command gives the implicitly defined derivative of an equation in which one variable is implicitly defined in terms of another variable. The syntax for this command is `impDif(Equation, Independent Variable, Dependent Variable)`. For example, the command `impDif(x²-x·y+y²=r²,x,y)` returns the result `(2x-y)/(x-2y)`.

> If a term such as xy is in the equation, be sure to separate the variables by including multiplication (⌨) so that TI-Nspire CAS doesn't think xy is its own variable.

Part III
The Graphs Application

The 5th Wave   By Rich Tennant

In this part . . .

This part gets into one of TI-Nspire's most powerful applications. You learn how this application is used to graph functions, inequalities, scatter plots, polar equations, parametric equations, differential equations and sequence graphing. I show you how to use tools to analyze your graphs to find critical points, as well as, how to use color to distinguish your graphs.

As with Part II, I dedicate the last chapter in this part to highlight some of the ways the computer algebra system functionality of TI-Nspire CAS can be used with the Graphs application.

Chapter 9

Working with Graphs

In This Chapter

▶ Graphing and analyzing functions

▶ Customizing your graph

▶ Exploring tools that allow you to get the most out of your graphs

▶ Working with other types of graphs

*I*n this chapter, I talk about the functionality built into TI-Nspire that enables you to graph and analyze functions, inequalities, parametric equations, polar equations, dynamic sequences, phase plots, differential equations, piecewise functions, and 3D graphing.

Graphing Functions

Press ctrl I ⇨Add Graphs to insert a new Graphs page into a current document. Or, open a new document and press menu ⇨Add Graphs.

By default, your new Graphs page opens with a coordinate graph displayed on the screen. Also, notice the blinking cursor located on the *entry line,* the narrow space located at the bottom of the screen. The entry line is used to type functions, type inequalities, configure scatter plots, and so on.

The equation for the area of a rectangle with a fixed perimeter of 12 units is area = $w(6 - w)$. The cursor is automatically located next to the first available function on the entry line. Here are the steps to type the equation:

1. **Start typing the equation, making sure that you press X for the independent variable.** The first screen in Figure 9-1 shows the complete equation on the entry line.

To type an equation with a set of parentheses, make sure that you include the appropriate operation symbol. In this case, after you press X, press × and then press (6 − X). If you forget to include the operation symbol, you may receive an error message (as pictured in the second screen of Figure 9-1).

2. **After typing the equation, press** ⌨enter⌨ **to activate the graph.**

Your equation in the third screen in Figure 9-1 appears in the *work area,* the space above the entry line. Notice that you cannot see all of the graph that you might want to see. I tell you how to adjust the window settings later in this chapter.

Figure 9-1:
Graphing a
function.

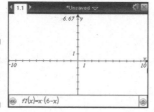 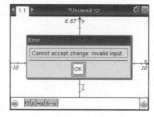 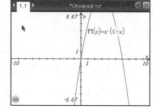

Moving from the entry line to the work area

After you press ⌨enter⌨ to graph an equation, a movable cursor appears in the work area and the entry line closes. If your cursor is in the entry line and you would like to move to the work area, you have a few options. To move to the work area, press ⌨esc⌨. Alternatively, press ⌨tab⌨ twice to move to the work area, which will make the entry line turn dark gray. To move back to the entry line, press ⌨tab⌨ once.

Hiding the entry line

If you hide the entry line, your work area will be slightly larger. Press ⌨tab⌨ to hide the entry line or press ⌨menu⌨⇨View⇨Hide Entry Line. To show the entry line, press ⌨menu⌨⇨View⇨Show Entry Line.

I love having the ability to toggle between Hide Entry Line and Show Entry Line by simply pressing the shortcut key sequence ⌨ctrl⌨ ⌨G⌨.

Adjusting the window settings manually

Referring to Figure 9-1, the window settings clearly do not effectively reveal the graph. The *x*-values of the function represent the width of the rectangle, and the *y*-values represent the area. The *y*-values are positive. Therefore, it makes sense to focus on the first quadrant only.

To grab the entire coordinate plane, move the cursor to some open space and press ⌷ctrl⌷⌷. You see the crunched up paper symbol with a hand, ✋, indicating that the entire coordinate plane can be translated in any direction. Use the Touchpad keys to move the graph so that the origin is located in the lower-left corner of the screen. Press ⌷esc⌷ to release the coordinate plane. See the first screen in Figure 9-2.

Unfortunately, the maximum point on the parabola is still not in view. Here's how to grab the tick marks on the axes themselves to change these end values:

1. **Move the cursor over the *x*-axis. Both axes start to pulse, and the word** *axes* **appears on the screen.**

   If another object (like text) gets in the way of finding the *x*-axis, press ⌷tab⌷ to cycle through and find any layered objects.

2. **Press ⌷ctrl⌷⌷ to grab both axes.**

3. **Press the Touchpad keys to drag the tick marks. Moving away from the origin zooms out. Moving toward the origin zooms in.**

4. **Press ⌷esc⌷ when the vertex of the parabola comes into view.**

Perhaps you want to adjust the scale of the *x*-axis only to completely fill the space in the work area with your graph. Follow these steps to do so:

1. **Move the cursor over the *x*-axis. Again, both axes start to pulse, and the word** *axes* **appears on the screen.**

2. **Press ⌷ctrl⌷⌷ to grab *both axes*.**

3. **Press and hold ⌷⇧shift⌷ as you press the Touchpad keys to drag the tick marks, moving toward or away from the origin to adjust the scale of only the *x*-axis.**

4. **Press ⌷esc⌷ to release the *x*-axis.**

The last two screens in Figure 9-2 show the result of dragging both axes and a single axis, respectively.

Figure 9-2:
Adjusting
the window
settings.

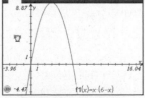
Translating the entire
coordinate plane

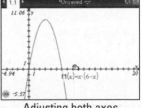
Adjusting both axes

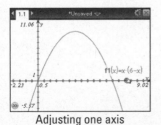
Adjusting one axis

Here are two more options that can be used to adjust the window settings:

✔ **Move the cursor to an axis end value (assuming that it's in view) and press [?] twice to highlight the current end value.** Type a new end value and press [enter] to put the change into effect. To edit all four end values one after the other, don't press [enter] after changing an end value. Instead, press [tab] to rotate in a clockwise manner to the next end value.

✔ **Press [menu]⇨Window/Zoom⇨Window Settings.** This action opens a dialog box and enables you to type your minimum and maximum end values as well as the scale. Press [enter] to put the changes into effect.

Using the zoom tools to set your window

Sometimes it is easier to automatically set your window on a Graphs page. There are eight different automatic zoom tools. To access the zoom tools, press [menu]⇨Window/Zoom and choose your tool:

✔ **Zoom User:** Try this one first. It saves your current window settings so that you can recall the settings after using the other zoom tools.

✔ **Zoom Standard:** This is the default setting.

✔ **Zoom Quadrant I:** Automatically sets your maximums and minimums to emphasize the first quadrant. See the first screen in Figure 9-3.

✔ **Zoom Trig:** This setting is great for trigonometric graphs. It automatically sets the x-min and x-max to integer multiples of π.

✔ **Zoom Data:** Recalculates the window so that all your data points are visible on the graph.

✔ **Zoom Fit:** Assures that the window includes all the minimum and maximum y-values. Be sure to set the appropriate x-values before using the Zoom Fit command. In the area function, the maximum is clearly visible, but the window includes much more than just the first quadrant (see the second screen in Figure 9-3).

✔ **Zoom Square:** Resizes the window so that your x- and y-axis scales are equal, which, on TI-Nspire, is a 3:2 ratio of the x-axis to the y-axis.

✔ **Zoom Decimal:** Recenters the origin and sets the minimums and maximums to a scale of 0.1. In this window, the Graph Trace tool automatically steps in tenths.

If you are not sure which Zoom tool you would like to use, there is a way to toggle from one Zoom tool to another. Hover your cursor over the axis, right-click ([ctrl] [menu]), and choose Attributes. Use your Touchpad arrow keys to adjust the second setting, and when you find a Zoom setting that you like, press [enter] to make the change.

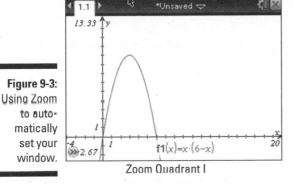
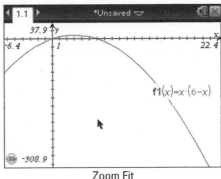

Zoom Quadrant I   Zoom Fit

Figure 9-3:
Using Zoom
to auto-
matically
set your
window.

Adjusting the scale manually

By default, the axes tic labels are shown on the coordinate plane in close proximity to the origin. These can be accessed by pressing [menu]⇨Window/ Zoom⇨Window Settings. Press [tab] to navigate to XScale. The default setting is Auto. While the word is highlighted, type the scale that you would like to set for the *x*-axis. Repeat for the *y*-axis, and press [enter] twice to put the change into effect.

The scale can also be changed manually on the Graphs page: Move the cursor to one of the axes tic labels (located near the origin). Press [↕] twice to edit the current tic label. Change the value for the tic label and press [enter] to put the change into effect.

The tic labels can be set to fractions. TI-Nspire CAS will even display irrational numbers like π. For example, if I graph $f(x)=\sin(x)$, I like to change the tic label on the *x*-axis to $\pi/2$. You may be surprised to find out that it will display as the exact fraction, $\pi/2$!

Changing your graph's attributes

TI-Nspire gives you the option of changing your graph's line weight (thin, medium, thick), line style (continuous, dotted, dashed), label style, and whether the graph is continuous or discrete. Press [menu]⇨Actions⇨Attributes, move your cursor over an object until an ⇔ appears, then click on the object. Alternatively, perform a right-click by moving your cursor to the graph and pressing [ctrl][menu]⇨Attributes.

After opening the Attributes menu, press the ▲▼ keys to move through the different options (line weight, line style, and so on). Use the ◆ keys to view the different options within each attribute and preview.

To put an attribute into effect, you must use the Touchpad keys to select it and then press ⏎. To exit the Attributes menu without making a change, press 🄴ᔕ𝖼.

Figure 9-4 shows some examples of different attributes that you can assign to a graph.

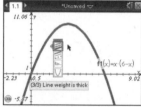 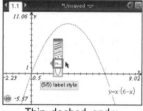 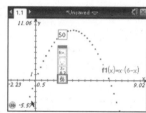

Figure 9-4:
Changing your graph's attributes.

Thick and continuous   Thin, dashed, and y= label style   Discrete and number of point

Changing the color

If you are using TI-Nspire CX or TI-Nspire Computer Software, then you can do more than just change the shades of grey, you can change the color of the graphed function. Each new function that you add in the Graphs application comes in a new color. To change the color of a graphed function, hover over the function and press 🄲𝗍𝗋 🗏menu ⇨Color. Click on the color of your choosing.

Adjusting the settings

To adjust the settings, press Menu⇨Settings. Changing the Display Digits field affects the precision of the points that are found in the Graphs environment. Notice that the Graphing Angle field (a graphed function) and the Geometry Angle field (a geometric construction) are considered different settings.

Analyzing your graph

Some of the more common analyses that are performed on a graphed function include evaluation, finding local maximum and minimum values, and locating zeros.

Using the Point On tool

The Point On tool offers a convenient way to accomplish each of these tasks. Here are the steps for using the Point On tool:

1. **Press menu⇨Geometry⇨Points & Lines⇨Point On to open the Point On tool.**

2. **Use the Touchpad to move the cursor and click on the graph.**

   A ghosted image of the coordinates at this location appears along with the words *point on*.

3. **Press enter or 🖰 to create the point.**

4. **Press esc to exit the Point On tool.**

Hovering over the tool icon in the upper-left corner of the screen gives tooltips for how to operate each tool on a Graphs page (see Figure 9-5).

Figure 9-5:
Using the Point On tool to analyze a graph.

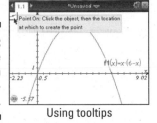
Using tooltips

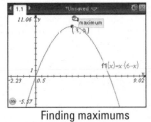
Finding maximums

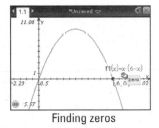
Finding zeros

Now move over the point until the ⌐ symbol appears, and press ctrl 🖰 to grab the point. Using the Touchpad, move your point along the graph. Notice that when you approach a local maximum value, the word *maximum* appears accompanied by the coordinates of this local extreme value. As you pass through a local minimum, you see the word *minimum*. As you pass through a zero (*x*-intercept) the word *zero* appears.

By default, the coordinate points are given with three significant digits. To change the level of precision, hover your cursor over the *x*-coordinate so that the ⌐ symbol appears and the coordinate blinks. Press the ⊞ key repeatedly to increase the number of displayed digits, and press the ⊟ key to decrease the number of displayed digits. Repeat this process for the *y*-coordinate.

As you hover the cursor over the *x*-coordinate, try clicking 🖰 twice to allow editing of the *x*-coordinate. Type a new *x*-coordinate value and press enter. Watch the point jump to its new location. You can also edit the *y*-coordinate. For functions that are not one to one (meaning in some instances that the *y*-values are not unique), TI-Nspire jumps to a point with the specified *y*-value closest to the current location.

Using the Trace tool

You also have the option of using the Trace tool to perform a similar analysis.

First, I'll change the function on the Graphs page to $f1(x) = (x^4 - 4x^3 + 3x^2) / (x - 1)$. A graph label can be edited on a graph page by moving the cursor over the label and pressing enter twice. If you are having trouble clicking on the graph label, press tab until the word *label* appears, then double-click. Follow these steps to use the Trace tool:

1. **Press del until your cursor gets to the = sign.**

2. **Type the equation. Press ctrl ÷ to use the fraction template.**

3. **Include exponents by following these steps:**

   a. *Press △ to move into Exponent mode.*

   b. *Type the exponent.*

   c. *Press ▶ to move out of Exponent mode.*

4. **Press enter to graph the function.**

   The first screen in Figure 9-6 shows the equation on the entry line.

Press menu ➪Trace➪Graph Trace to invoke this tool. A movable point and its coordinates are automatically placed on the graph. Use the ◀▶ keys to move this point along the graph. You can periodically place points on your graph by pressing enter. (See the second screen in Figure 9-6).

The label on a graph is "magnetized" to the graph of the function. If the label gets in the way, move the cursor over the label. Press ctrl 🔒 and move the label to a more convenient location. See the second screen in Figure 9-6. Press esc to place the label in its new location. Right-clicking the label by pressing ctrl menu also gives you a choice to hide the label of the function.

While using the Graph Trace tool, you can jump to values by typing the *x*-value that you would like to trace. For example, a hole exists in this graph at $(1,-2)$. Press 1 enter to find the hole on the graph. The third screen of Figure 9-6 shows the discontinuity with a dotted vertical line. The ordered pair reads (1,undef).

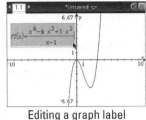
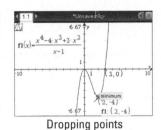
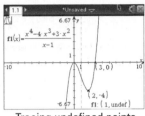

Figure 9-6:
Using the
Graph Trace
tool.

| Editing a graph label | Dropping points | Tracing undefined points |

Using the Analyze Graph tool

Another way of detecting points of interest on a Graphs page is to use
the Analyze Graph tool. Use the tool to find a local maximum by pressing
[menu]⇨Analyze Graph⇨Maximum. The on-screen prompt asks, "lower bound?"
Move the cursor and press [enter] or enter a number and press [enter] to choose
the lower bound of the search region. Another prompt asks, "upper bound?"
Again, move the cursor and press [enter] or enter a number and press [enter] to
choose the upper bound of the search region. The tool identifies the local
maximum and plots the ordered pair (see Figure 9-7). This tool can also be
accessed by right-clicking ([ctrl] [menu]) the graph of the function.

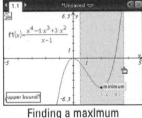

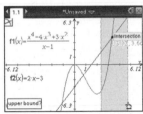
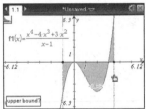

Figure 9-7:
Using the
Analyze
Graph tool.

| Finding a maximum | Finding an intersection point | Finding the area under a curve |

Open the entry line by pressing [ctrl] [G]. Enter the function f2(x) = $2x - 3$.

The Analyze Graph tool can also be used to find points of intersection; however, it can find only one point of intersection at a time. The smart thing to
do is to only include the point of intersection that you want in the specified
range of the function graph. In the second screen of Figure 9-7, I used this
technique to find one of the intersection points of f1(x) and f2(x).

In addition, the Analyze Graph tool can find the derivative at a point and the
definite integral. (See the third screen of Figure 9-7.)

TIP

Using TI-Nspire CAS, you can use the Analyze Graph tool to find an inflection
point as long as there are no discontinuities in the selected region.

Hiding or deleting a graph

To hide a graph without actually deleting it, move your cursor to the graph and right-click by pressing [ctrl] [menu]⇨Hide.

To reveal a hidden graph, you must press [menu]⇨Actions⇨Hide/Show to open the Hide/Show tool. A ghosted image of all hidden objects appears, including your hidden graph.

To delete a graph, move your cursor to the graph and right-click by pressing [ctrl] [menu]⇨Delete.

If you mistakenly delete a graph, press [ctrl] [esc] or [ctrl] [Z] to undo.

Graphing an f(y) function

In this section, I show you a method of graphing an f(y) function, in other words, a function that has been solved for x instead of y. Before doing so, I recommend pressing [ctrl] [G] to hide the entry line.

Here is an alternative method that I use to graph the function $x = y^2 - 2$:

1. **Move your cursor to some open space and right-click by pressing** [ctrl] [menu]⇨**Text to open the Text tool.**

2. **Type** $x = y^2 - 2$ **and press** [enter] **to close the text box.**

3. **Press** [esc] **to close the Text tool.**

4. **Grab (press** [ctrl] [⬚]**) and drag the text expression to the x-axis and notice that a ghosted image of the graph appears. Press** [enter] **to graph this function.**

   You should approach the x-axis very slowly for this method to work. The sequence of screens in Figure 9-8 shows how this method works.

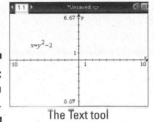
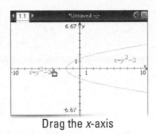
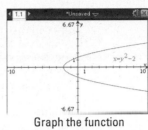

Figure 9-8:
Graphing an
f(y) function.

The Text tool   Drag the *x*-axis   Graph the function

Analyzing two functions simultaneously

The Trace All tool enables you to analyze two functions simultaneously.
Open the entry line (ctrl G), type the function $y = x+1$, and press enter. Press
menu↪Trace↪Trace All to start tracing along both functions simultaneously.
The Trace All tool does not allow you to recognize points of interest as
you move along the graphs. While using the Trace All tool, you may enter a
number (and then press enter) so that the cursor will jump to that independent value. See Figure 9-9.

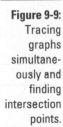
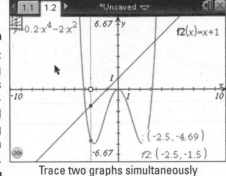
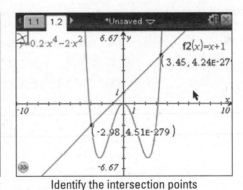

Figure 9-9:
Tracing
graphs
simultane-
ously and
finding
intersection
points.

Trace two graphs simultaneously   Identify the intersection points

I want students to be able to find the intersection point(s) of two functions as quickly as they can. The best tool to accomplish this task is the Intersection Point(s) tool. My favorite feature of this tool is that it plots the points and labels the coordinates. Here are the steps to find the intersection points:

1. **Press** [menu]⇨**Geometry**⇨**Points & Lines**⇨**Intersection Point(s).**

2. **Move your cursor over each function and press** [🔒].

3. **Press** [esc] **to exit the Intersection Point(s) tool.**

Adding a tangent line to a graph

In this section, I start with a fresh Graphs page and graph the equation $f(x) = x^3 - 3x^2 - 2x + 6$.

The Tangent Line tool is a great option for exploring the instantaneous rate of change of a nonlinear function. Here's how it works:

1. **Graph your function on the first available function on the entry line.**

2. **Adjust the window to reveal all your points of interest.**

3. **Press** [menu]⇨**Geometry**⇨**Points & Lines**⇨**Tangent to open the Tangent Line tool.**

4. **Click** [🔒] **on the function graph, then press** [🔒] **again to construct the tangent line.**

5. **Press** [esc] **to exit the Tangent Line tool.**

The second screen in Figure 9-10 shows the result of these steps. You can drag the point of tangency to move the tangent line along the graph.

The third screen in Figure 9-10 shows an alternative approach to this construction. Here, I've constructed a line perpendicular to the x-axis. I then constructed the point of intersection of this perpendicular line with the function. Then, I constructed the tangent to this point of intersection. Under this scenario, I can move the tangent line by dragging the point of intersection of the x-axis and the perpendicular line. I like this option a bit more because the tangent line moves along the curve more smoothly. Finally, I pressed [menu]⇨Measurement to measure the slope of the tangent line.

Figure 9-10:
Adding a
tangent line
to a graph.

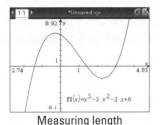
Measuring length

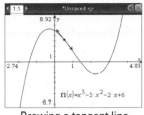
Drawing a tangent line

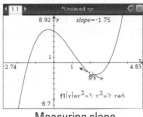
Measuring slope

Using Measurement Transfer

The Measurement Transfer tools allow you to transfer the result of a calculation, a numeric text value, or a measurement to a circle, ray, or vector. You
can also use this tool to transfer a measurement to an axis, which is what I
do in this section. Specifically, I'm interested in transferring my slope of the
tangent line to the y-axis. Here are the steps:

1. **Press menu⇨Geometry⇨Construction⇨Measurement Transfer to invoke
   the Measurement Transfer tool.**

2. **Move the cursor to the result of the slope measurement and press** 🔳**.**

3. **Move the cursor to the y-axis and press** 🔳**.**

4. **Press esc to exit the Measurement Transfer tool.**

Notice that a point appears on the y-axis with a y-coordinate equal to the current value of the slope calculation. Try dragging the hollow point on the tangent line and watch as the point on the y-axis moves up and down.

Figure 9-11 shows the results of using the Measurement Transfer tool. I've
adjusted the y-axis to provide a better window for seeing this result. The
first two screens show the movement of the transferred measurement as the
x-axis point is dragged. In the third screen, I've added perpendicular lines
through the transferred measurement and hollow point on the tangent line.
Press menu⇨Geometry⇨Construction⇨Perpendicular to access the perpendicular line tool. Then, I found the intersection point of the two perpendicular
lines.

Figure 9-11:
The
Measure-
ment
Transfer
tool.

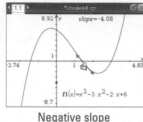 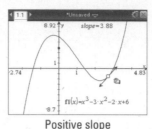 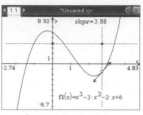

Negative slope   Positive slope   Perpendicular lines

Using the Locus tool and Geometry Trace

The intersection point of the perpendicular lines on my graph is now part of a new function. The Locus tool and Geometry Trace provide two options for "seeing" the entire function given by the path of this point as the hollow point on the tangent line is dragged.

Here's how to use the Locus tool to investigate this new function:

1. **Press** menu⇨**Geometry**⇨**Construction**⇨**Locus to invoke the Locus tool.**

2. **Move the cursor to the intersection point of the perpendicular lines (the point whose path I'm interested in viewing) and press** 🔘**.**

3. **Move the cursor to the hollow point on the tangent line, the driver point, and press** 🔘**.**

4. **Press** esc **to exit the Locus tool.**

The first screen in Figure 9-12 shows the result of using the Locus tool. In the second screen in Figure 9-12, I've changed its attributes to dashed and changed its thickness to medium to provide some contrast from the other graphed function.

Geometry Trace offers another way to see this relationship. Hover over the locus function, press ctrl menu⇨Hide to hide the locus, and follow these steps:

1. **Press** menu⇨**Trace**⇨**Geometry Trace to invoke the Geometry Trace tool.**

2. **Move the cursor to the intersection point of the perpendicular lines (the point whose path I'm interested in viewing) and press**🔘**.**

3. **Move the cursor to the hollow point on the tangent line, the driver point, and press** ctrl🔘 **to grab the point.**

4. **Use the Touchpad keys to move the grabbed point along the function and watch the trace appear on the screen.**

5. **Press** esc **to exit the Geometry Trace tool.**

This is a nice way to see the graph of the derivative of a function. I like using the Geometry Trace tool so that I can control how fast the derivative is formed. The quicker I drag the point on the *x*-axis, the fewer points are left behind in the trace track. Incidentally, a limit of 150 points is left in the track. The older parts of the track begin to fade out as you continue to track the movement of the point. See the third screen in Figure 9-12.

Figure 9-12:
Using the
Locus tool
and the
Geometry
Trace tool.

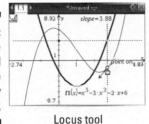 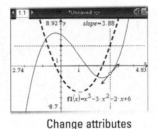 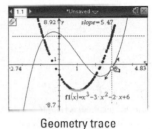

Locus tool   Change attributes   Geometry trace

TIP

Press menu⇨Trace⇨Erase Geometry Trace to erase the Geometry Trace.

Adding a function table

To add a function table to a graph, press menu⇨View⇨Split-screen⇨Table. Or, use the shortcut and press ctrl T. This action automatically splits the screen and adds a Lists & Spreadsheet application with, by default, *x* values incrementing by 1 and their corresponding *y*-values. Use the ▲▼ keys to scroll through the function table. See the first screen in Figure 9-13.

Figure 9-13:
Adding
multiple
function
tables on
a List &
Spread-
sheet page.

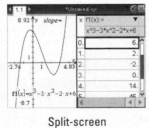

Split-screen   Lists & Spreadsheet   Comparing functions

You have the option to change the table settings. Press menu⇨Table⇨Edit Table Settings to open a dialog box and customize Table Start, Table Step, and the settings for the Independent and Dependent variables.

Adding a function table changes the page layout from one application to two applications. Press ⌜ctrl⌝⌜T⌝ to delete the table and return to a full-screen view of the Graphs page.

If you have more than one function graph, you need to add a List & Spreadsheet page to display two (or more) function tables simultaneously. Follow these steps:

1. **Insert a new Lists & Spreadsheet page. Press ⌜ctrl⌝⌜doc▾⌝⇨Add List & Spreadsheet.**

2. **Use the shortcut to add a function table. Press ⌜ctrl⌝⌜T⌝⇨f2.**

   The second screen in Figure 9-13 shows the result of this step. For the next step, I needed a second function, so I went back to the graphs page and added $f2(x) = x + 1$.

3. **Move your cursor to the down arrow in the upper-right corner of the screen. Press ⌜⌗⌝ and choose a function for your second function table.**

   I chose f2 as my second function in the table. See the third screen in Figure 9-13.

My favorite way of adding (or deleting) a table is to press ⌜ctrl⌝⌜T⌝.

Other investigations

So far, I've showcased many of the tools used within the Graphs application. The following sections introduce a few other noteworthy tools and features.

Graphing transformations

Several types of functions have graphs that can be directly manipulated on the screen. To accomplish this task, simply press ⌜ctrl⌝⌜⌗⌝ to grab the graph and then use the Touchpad keys to perform a transformation. As an example, create the graph of $y = x^2$ as shown in the first screen in Figure 9-14. Two different options are possible:

✔ **Perform a translation. Position the cursor on the vertex of the graph until the ✛ symbol appears, and press ⌜ctrl⌝⌜⌗⌝ to grab the graph. Use the Touchpad keys to translate the graph and press ⌜esc⌝ when finished.**

   Notice that the equation of the graph is updated automatically, in real time, as you move the graph, as shown in the second screen in Figure 9-14.

✔ **Perform a stretch. Position the cursor on a side of the parabola until the ⤢ symbol appears and press ⌜ctrl⌝⌜⌗⌝ to grab the graph. Use the Touchpad keys to stretch the graph and press ⌜esc⌝ when complete.**

   Notice that the value a in front of the parentheses, is automatically updated, as shown in the third screen in Figure 9-14.

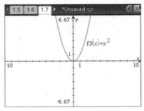 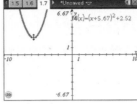 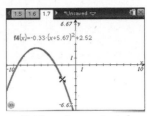

Figure 9-14:
Performing
transforma-
tions on
the graph
of $y = x^2$.

Graph a quadratic   Performing a translation   Performing a stretch

Here is a list of the different functions that can be transformed using the same procedures just described:

- Linear functions of the form $y = b$, where b is a constant

- Linear functions of the form $y = ax + b$, where a and b are constants

- Quadratic functions of the form $y = ax^2 + bx + c$, where a, b, and c are constants or the form $y = a(x - h)^2 + k$

- Exponential functions of the form $y = e^{ax + b} + c$, where a, b, and c are constants

- Exponential functions of the form $y = be^{ax} + c$, where a, b, and c are constants

- Exponential functions of the form $y = de^{ax + b} + c$, where a, b, c, and d are constants

- Logarithmic functions of the form $y = a \ln(cx + b) + d$, where a, b, c, and d are constants

- Sinusoidal functions of the form $y = a \sin(cx + b) + d$, where a, b, c, and d are constants

- Cosinusoidal functions of the form $y = a \cos(cx + b) + d$, where a, b, c, and d are constants

TIP

With a bit of practice, you will quickly learn where to find the positions on the graph where the ✛ (translation) and ✕ (stretch) symbols appear. For example, translate the graph of $a \sin(cx + b) + d$ by positioning the cursor at a point halfway between the maximum and minimum values. Any other point on the graph allows you to stretch the graph.

Using a slider

In the previous section, I mention that $y = e^{ax + b} + c$ is one type of function that can be transformed. What if I want to transform an exponential function with a base other than e? Better yet, what if I want to explore the graph of $y = ab^x$, where a and b can take on any range of values? To perform this investigation, use the slider feature and follow these steps:

1. **Press [menu]⇨Actions⇨Insert Slider. Move the ghosted slider to the desired location and press 🔲.**

   This action inserts a slider box on the screen. By default, the slider values range from 0 to 10, with an initial value of 5 and step size of 1. Move your cursor over the range values and press 🔲 to change them. Notice that a grayed out $v1$ is displayed in the upper-left corner of the slider box. This is where you name your slider variable in the next step.

2. **Type a variable name (using standard variable-naming conventions) and press [enter].**

   I typed a for my variable name, as shown in the first screen in Figure 9-15.

3. **Add additional sliders as needed.**

   I have added a second slider for the constant b in the function $y = ab^x$.

4. **Type your function in the entry line using the variables defined in your sliders.**

   You can also access these variables by pressing the [var] key.

5. **Click 🔲 the up or down arrows of the slider to change the value.**

   Watch your graph update automatically, as demonstrated in the second screen in Figure 9-15.

To save time, I like to position the cursor on the slider bar and press 🔲. Then, I can click the Touchpad keys to operate the slider (without ever having to grab!). This technique also works to operate points placed on a grid.

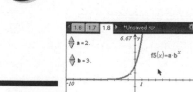

Figure 9-15:
Working
with sliders.

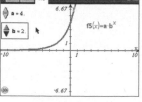

Using sliders   Clicking sliders   Slider settings

Try moving your cursor to the top-right of the slider box and performing a right-click by pressing ⌈ctrl⌉⌈menu⌉ to access the context menu. This gives you the option of changing the slider settings, as shown in the third screen in Figure 9-15. You also have the option of animating the slider. This action moves the slider back and forth between the minimum and maximum slider values. Access the slider context menu a second time to stop the animation.

Graphing Inequalities

With TI-Nspire, you can graph a single inequality or multiple inequalities. To graph an inequality, follow these steps:

1. **Press ⌈ctrl⌉⌈G⌉ to open the entry line and position the cursor to the right of the first available function.**

2. **Press ⌈del⌉ until the equals sign is deleted.**

   See the first screen in Figure 9-16.

3. **Type a number to choose your inequality, enter the inequality, and press ⌈enter⌉ to view its graph.**

   See the second screen in Figure 9-16

4. **Repeat Steps 1 through 3 to graph additional inequalities.**

   See Figure 9-16.

Figure 9-16:
Graphing
inequalities.

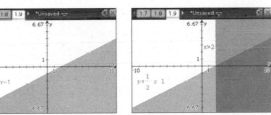
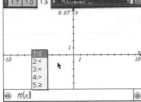

Choose inequality sign   Graph inequality   Graph an f(*y*) inequality

Here are few additional things I'd like to point out about inequality graphing:

✔ **You have the option of using the Text tool to enter an f(*y*) inequality.** Press ⌈ctrl⌉⌈menu⌉ in an open space and select Text. Enter an f(*y*) inequality (I entered *x* > 2), then press ⌈enter⌉. Press ⌈ctrl⌉⌈🖰⌉ to grab the text box and use the Touchpad keys to drag and drop it on the *x*-axis. The result is shown in the third screen in Figure 9-16.

✔ **Less than (<) and greater than (>) inequalities have dashed boundary lines.**

✔ **Less than or equal to (≤) and greater than or equal to (≥) inequalities have solid boundary lines.**

✔ **Use the Intersection Point(s) tool (press [menu]⇨Geometry⇨Points & Lines⇨Intersection Points) to find the intersection point of your inequalities.**

Change the fill color of the inequality shading to better see the solution region. Did you know that blue and red make purple when the shaded regions overlap? To change the color of the shaded regions, right-click by pressing [ctrl] [menu]⇨Color⇨Fill Color.

Graphing Parametric Equations

Press [menu]⇨Graph Type⇨Parametric to switch to parametric graphing mode. Alternatively, move to the entry line and press [ctrl] [menu]⇨Parametric. Next, follow these steps:

1. **Type the x-component equation, using t as the independent variable.**

   TI-Nspire uses the notation $x1(t)$ for the first x-component, $x2(t)$ for the second x-component, and so on.

2. **Type the y-component equation, using t as the independent variable.**

   TI-Nspire uses the notation $y1(t)$ for the first y-component, $y2(t)$ for the second y-component, and so on.

3. **Edit the interval for the variable t and the *tstep* increment.**

   By default, parametric graphing is configured in radians with $0 \le t \le 2\pi$ and $tstep = \pi/24$.

4. **Press [enter] to graph the parametric curve.**

Figure 9-17 shows an example of a parametric graph.

Figure 9-17: Graphing parametric equations.

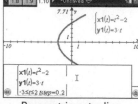
Parametric entry line

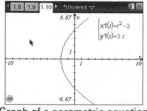
Graph of a parametric equation

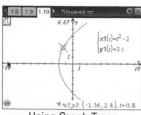
Using Graph Trace

In the second screen in Figure 9-17, I pressed ⌈ctrl⌉⌈G⌉ to hide the entry line.

In the third screen in Figure 9-17, I used Graph Trace (press ⌈menu⌉⇨Trace⇨ Graph Trace) to trace along the graph. Each time I press ◄ or ►⌈,⌉ the trace moves by a value of *t* equal to *t*-step.

Graphing Polar Equations

Press ⌈menu⌉⇨Graph Type⇨Polar to switch to polar graphing mode. Alternatively, move to the entry line and press ⌈ctrl⌉⌈menu⌉⇨Polar. Next, follow these steps:

1. **Type the equation, using θ as the independent variable.**

   Press the ⌈π·⌉ key to access the θ symbol.

2. **Type the interval for the variable θ and the θ *step* increment.**

   By default, polar graphing is configured in radians with $0 \le \theta \le 2\pi$ and θ *step* = π/24.

3. **Press ⌈enter⌉ to graph the polar curve.**

Figure 9-18 shows an example of a polar graph.

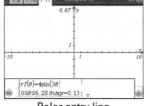 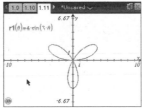 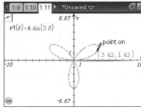

Figure 9-18: Graphing polar equations.

| Polar entry line | Graph of a polar equation | Using the Point On tool |

In the third screen in Figure 9-18, I use the Point On tool (press ⌈menu⌉⇨Geometry⇨Points & Lines⇨Point On) to trace along the graph.

As with parametric graphing, many of the features of function graphing are available in polar graphing mode.

Graphing Sequences & Phase Plots

To demonstrate the dynamic nature of sequence graphs, I will enter two sequences that represent the population of rabbits and foxes. Of course, the population of predators is related to the population of its prey. Press menu⇨Graph Type⇨Sequence⇨Sequence to enter the sequence for rabbits: u1=u1(n–1)·(1.05–.001·u2(n–1), initial condition=200, 0<n<500. See the first screen in Figure 9-19. In the sequence for foxes, u2 = u2(n–1)·(0.97+.0002·u1(n–1), initial condition=50, 0<n<500. See the graph of both sequences in the second screen in Figure 9-19.

If I want to see how one sequence affects another sequence, I can use a phase plot to represent the data. Here are the steps to create a phase plot:

1. **Press** ctrl I ⇨**Add Graphs.**
2. **Press** menu⇨**Graph Type**⇨**Sequence**⇨**Custom.**
3. **Enter** *u1* **for** *x* **and** *u2* **for** *y* **with** 1<*n*<300.
4. **Press** enter **to graph the phase plot.**
5. **Adjust your window by pressing** menu⇨**Window/Zoom**⇨**Zoom Fit.**

   See the third screen in Figure 9-19.

You can grab and move the point that represents the initial condition right on the graph screen.

Figure 9-19:
Graphing
Sequences
& Phase
plots.

Sequence equation

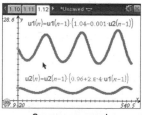
Sequence graph

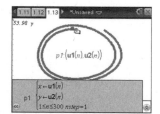
Phase plot graph

Graphing Differential Equations

Graphing differential equations is another new feature on TI-Nspire. You can set the initial condition(s), customize the slope field, and choose your solution method (Euler or Runge-Kutta). Oh yeah, and you can grab the initial condition and change it right on the graph screen. Follow these steps to graph a differential equation:

1. **Press** menu⇨**Graph Type**⇨**Diff Eq.**

2. **Type the differential equation, $y1 = 0.2x^2$.**

   The default identifier is $y1$. To change the identifier, click the box to the left of the entry line. You may reference the identifier in the entry line.

3. **Enter an initial condition, (0,0), as an ordered pair.**

   Press the Add Initial Conditions button to enter up to three additional initial conditions for each differential equation.

4. **Press the Edit Parameters button (...) if you would like to customize the differential equation any further.**

5. **Press** enter **to graph the differential equation or press the down arrow to display the next differential equation edit field.**

   See the first screen in Figure 9-20.

Entry line

Edit Parameters button

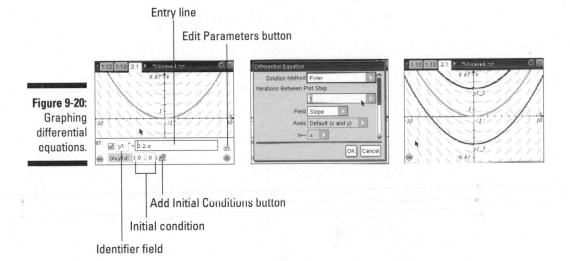

Figure 9-20:
Graphing
differential
equations.

Add Initial Conditions button

Initial condition

Identifier field

Differential equation settings can be accessed by pressing the Edit Parameters button (...). See part of the dialog box in the second screen in Figure 9-20. Here is a brief summary of the settings:

✔ **Solution Method:** You have a choice of using Euler or Runge-Kutta as the numerical solution method.

✔ **Iterations between plot step:** If you are using the Euler solution method, type an integer to set the computational accuracy.

✓ **Error Tolerance:** If you are using the Runge-Kutta solution method, type a number that is greater than or equal to 1×10^{-14}.

✓ **Field:** *None*— No field is plotted. *Slope*— Plots a slope field representing the solutions. *Direction*— Graphs a slope field representing the relationship between the values of two differential equations.

✓ **Axes:** *Default(x and y)*— Plots the x on the x-axis and the y on the y-axis. *Custom*— This setting lets you select the values to be plotted on each axis.

✓ **Plot Start:** This field determines the independent variable value where the solution plot starts.

✓ **Plot End:** This field determines the independent variable value where the solution plot stops.

✓ **Plot Step:** This field determines the increment of the independent variable where the values are plotted.

✓ **Field Resolution:** This field determines the number of columns used to draw the slope field. You can only change this field if Field=Direction or Slope.

✓ **Direction Field at x=:** This field sets the values of the independent variable at which a direction field is drawn when plotting autonomous equations (those that refer to x). You can only change this parameter if Field=Direction.

3D Graphing

Graphing in 3D helps students to visualize what a function would look like in 3D (and you don't need to wear uncomfortable glasses to see it). 3D Graphing allows you to enter an expression of the form $z(x,y)$. To graph and explore in 3D, follow these steps:

1. **Press [doc⌄]⇨Insert⇨Graphs.**

2. **Press [menu]⇨View⇨3D Graphing.**

   2D and 3D graphing are completely separate environments. In the first screen in Figure 9-21, you may notice the menu options are only those that are unique to the 3D graphing environment.

3. **Type a function in the entry line, $z1(x,y)=\sin(x)\cdot\cos(y)$ and press [enter] to graph it.**

   See the second screen in Figure 9-21.

4. **Use your Touchpad keys to rotate the graph.**

   For fun, press Ⓐ to auto-rotate the graph; pressing Ⓡ allows you to manually rotate the graph again using your Touchpad arrow keys. Press ÷ to shrink the box or × to magnify the box.

5. **Press** menu⇨**Trace**⇨**zTrace.**

   Hold the ⇧shift key down and use your Touchpad arrow keys to trace the graph. See the third screen in Figure 9-21.

6. **Explore some new color options of your 3D graph by right-clicking on the graph,** ctrl menu⇨**Color**⇨**Custom Plot Color.**

   You can vary the color of your graph by steepness or height. The top and bottom of your graph can be different colors.

Hover over the graph and press ctrl menu⇨Attributes to customize the transparency and resolution of your 3D graph.

3D graphing has many customizable features. You can graph multiple 3D graphs on the same axes. Take some time to explore this powerful feature — the possibilities are endless!

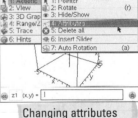 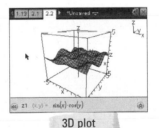 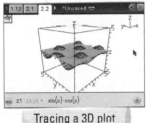

Figure 9-21: 3D graphing.

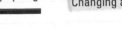

Changing attributes   3D plot   Tracing a 3D plot

Limit the number of 3D graphs that you put on a handheld. It takes a lot of computing power to draw 3D graphs, causing a memory drain. If you have too many 3D graphs in one document, your handheld may turn off and reboot.

3D Parametric Graphing

Why not have the best of both worlds by combining parametric graphing and the 3D capabilities of TI-Nspire? 3D parametric graphs can be used to plot lines, planes, spirals, and spheres just to name a few. Follow these steps to enter a 3D parametric graph:

1. **Press** docᵥ⇨**Insert**⇨**Graphs.**

2. **Press** menu⇨**View**⇨**3D Graphing.**

3. **Press** menu⇨**3D Graph Entry/Edit**⇨**Parametric.**

   See the first screen in Figure 9-22.

4. **Type a function in the entry line,** $xp1(t,u)=5\sin(u)$ enter $\cos(t)$, $yp1(t,u)=5\sin(u)$ enter $\sin(t)$, **and** $zp1(t,u)=5\cos(u)$.

   See the second screen in Figure 9-22.

5. **Press** enter **to graph.**

   See the third screen in Figure 9-22.

Figure 9-22: 3D parametric graphing.

 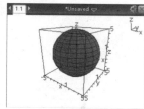

Press Menu   Enter functions   Press Enter to graph

Graphing Piecewise Functions

Press menu⇨Graph Type⇨Function to switch back to function graphing mode. Alternatively, move to the entry line and press ctrl menu⇨Function. Two different piecewise function templates are available. The Piecewise template (2-piece) allows you to create expressions and conditions for two restricted functions. I prefer the Piecewise template (N-piece), because you can dictate how many pieces to include in the template. To invoke the template, follow these steps:

1. **Press** ^ **to access the templates. Choose the Piecewise template (N-piece).**

   See the first screen in Figure 9-23.

2. **Type the number of function pieces when prompted. Press** enter.

   You have a limit of 50 piecewise functions using the template. (Why in the world would you ever need that many?) See the second screen in Figure 9-23.

3. **Edit the expression and condition for each piece.**

   Press tab or use the Touchpad keys to navigate the template.

4. **Press** enter **to graph the piecewise function.**

The third screen in Figure 9-23 shows an example of a piecewise function graph.

 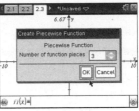 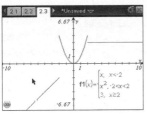

 Figure 9-23: Graphing piecewise functions.

Press Menu   Enter functions   Press Enter to graph

 To delete a function piece from an already-existing piecewise function, place the cursor at the start of the expression you would like to delete and press ⌈del⌋. On a PC (using TI-Nspire Teacher Software), press Alt⇨Enter to add a function piece to an already existing piecewise function.

 The conditions on each function piece may not intersect (or it wouldn't be a function). If you try to graph these "overlapping" pieces, the TI-Nspire will only graph the first function piece that you entered into the template.

Graphing an f(y) Piecewise Function

Only a select few TI-Nspire users know that you can graph an f(y) piecewise function where you restrict y instead of x. If you follow these directions, you can join this elite group of TI-Nspire users:

1. **Press ⌈doc▾⌋⇨Insert⇨Graphs.**

2. **Move your cursor to some open space and right-click by pressing ⌈ctrl⌋ ⌈menu⌋⇨Text to open the Text tool.**

3. **Type $x = (y+2)^2 + 1$, press ⌈ctrl⌋= and choose the Such That vertical bar.**

   See the first screen in Figure 9-24.

4. **Enter the restriction for the y-values and press ⌈enter⌋.**

   See the second screen in Figure 9-24.

5. **Grab (press ⌈ctrl⌋⌈✋⌋) and drag the text box to the x-axis and notice that a ghosted image of the graph appears. Press ⌈enter⌋ to graph this f(y) function.**

   See the third screen in Figure 9-24.

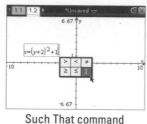
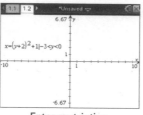

Figure 9-24:
Graphing
an f(y)
piecewise
function.

Such That command   Enter restriction   Drag and drop

Graphing Inverse Functions

Now that you know how to graph f(y) functions, graphing an inverse function will be a piece of cake. Follow these steps to graph an inverse function:

1. **Press** [doc▾]⇨**Insert**⇨**Graphs.**

2. **Enter a function in f1(*x*) and press** [enter] **to graph.**

   I entered **f1(*x*) = sin(*x*) + 2.** See the first screen in Figure 9-25.

3. **Move your cursor to some open space and right-click by pressing** [ctrl] [menu]⇨**Text to open the Text tool.**

4. **Enter** *x* = f1(*y*), **press** [enter] **to close the text box.**

   See the second screen in Figure 9-25.

5. **Grab (press** [ctrl][✋]**) and drag the text box to the *x*-axis and notice that a ghosted image of the graph appears. Press** [enter] **to graph this f(*y*) function.**

   See the third screen in Figure 9-25.

The inverse of the function, f1(*x*) = sin(*x*) + 2 is not a function. TI-Nspire graphs the inverse equation even if the inverse equation is not a function.

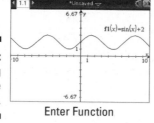
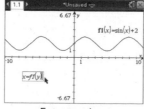
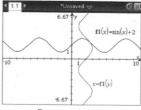

Figure 9-25:
Graphing
an inverse
function.

Enter Function   Enter text box   Drag and drop

Graphing Conics

TI-Npire has built-in conic graphing templates that allow you to quickly graph circles, ellipses, f(x) parabolas, f(y) parabolas and hyperbolas. Here are the steps to graph an ellipse:

1. **Press** doc⌄⤷**Insert**⤷**Graphs.**

2. **Press** menu⤷**Graph Entry/Edit**⤷**Equation**⤷**Ellipse.**

   See the first screen in Figure 9-26.

3. **Enter the missing values in the ellipse template.**

   See the second screen in Figure 9-26.

4. **Press** enter **to graph the ellipse.**

   See the third screen in Figure 9-26.

Have fun exploring all of the different templates. Press menu⤷Graph Entry/Edit⤷Equation to see all of the possibilities. If you choose Conic, you are able to enter an equation that graphs a tilted ellipse!

Hidden in the conic graphing templates are three linear templates. Press menu⤷Graph Entry/Edit⤷Equation⤷Line to access templates for slope-intercept form ($y = mx + b$), vertical lines ($x = c$), and standard linear form ($ax + by = c$).

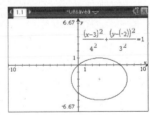

Figure 9-26: Graphing an ellipse.

Ellipse template   Enter missing values   Press Enter

Analyzing Conics

To analyze a conic, press [menu]⇨Analyze Graph⇨Analyze Conics and then choose one of these menu choices: Center, Vertices, Foci, Axes of Symmetry, Directrix, Asymptotes, Radius, Eccentricity, and the Latus Rectum. After choosing from the menu, simply click the graph of the conic to analyze the conic.

Chapter 10

Using the Graphs Application to Do Calculus

In This Chapter

▶ Understanding which features can be used in the Graphs application to do calculus

▶ Using TI-Nspire and TI-Nspire CAS tools to explore calculus

1n this chapter, I talk about some of the features that can be called into play in the Graphs application and describe how you might want to use them to explore calculus. Some of the tools are CAS-specific, which I have clearly labeled in the TI-Nspire CAS section.

Graphing Derivatives

The study of calculus should include a focus on the four key mathematical representations: algebraic, geometric, numeric, and verbal. The Calculator application takes care of the algebraic piece, the Lists & Spreadsheet application takes care of the numeric piece, and the Notes application can take care of the verbal piece. The Graphs application stands ready to take care of the geometric representation of calculus concepts.

To graph the function $y = x^4 - 8x^2 + 5$ and its derivative on the same screen, follow these steps:

1. **If you haven't already done so, open a new Graphs page.**

2. **Graph $y = x^4 - 8x^2 + 5$ using the first available function on the entry line.**

3. **Press** ⌃G **to open the entry line, then press** 🔢 **to open the Math template, highlight the derivative template, and press** ⏎.

   See the first screen in Figure 10-1. You may want to resize the window.

4. **Type** x **for the first field of the derivative template, and press** `tab` **to move to the second field.**

5. **Type** f1(x) **or press** `var` **to see a list of available variables, select f1 from the list, and press** `X` **to complete the expression f1(x). Press** `enter` **to graph the derivative.**

6. **Move the cursor to the graph of the derivative and press** `ctrl` `menu`⇨**Attributes. Change the line style to medium and dashed to provide some visual clarity between the two graphs.**

See the second and third screens in Figure 10-1. If you are using TI-Nspire CX, you may want to skip this step since each graph is a different color.

Figure 10-1:
Graphing
derivatives
with the
derivative
template.

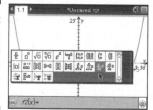 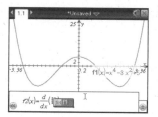 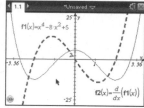

Try changing f1(x). The graph of f1(x) and its derivative updates accordingly.

Graphing Antiderivatives

Keep in mind that a function has an infinite number of antiderivatives. In the example given in the following section, I look at a particular antiderivative and then show you how to use a slider to investigate an entire family of curves defined by an antiderivative.

Using the definite integral template

To graph the antiderivative of $y = x^3 - 3x^2 - 2x + 6$, follow these steps:

1. **Press** `ctrl` `G` **to open the entry line, then graph** $y = x^3 - 3x^2 - 2x + 6$.

2. **Press** `M⦚` **to open the Math template, highlight the definite integral template, and press** `enter`.

See the first screen in Figure 10-2.

3. Press ⓪ to input the lower limit of the definite integral template, and then press ⟨tab⟩ to move to the upper-limit field. Press ⟨X⟩ and then press ⟨tab⟩ to move to the integral field. Type f1(x) or press ⟨var⟩ and select f1 from the list of variables, then press ⟨tab⟩ again to move to the next field of the definite integral template.

4. Type x in the last field and press ⟨enter⟩ to graph the antiderivative.

It may take a few seconds for the graph to form on a handheld.

Figure 10-2:
Graphing antiderivatives using the definite integral template.

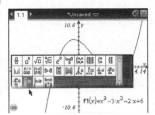 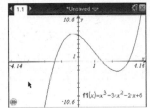 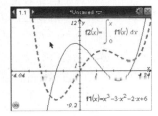

The antiderivative that is graphed in Figure 10-3 is defined by the equation

$$y = \frac{1}{4}x^4 - x^3 - x^2 - 6x$$

This equation is based on the general solution

$$y = \frac{1}{4}x^4 - x^3 - x^2 - 6x + C$$

with $C = 0$.

Using the indefinite integral template on TI-Nspire CAS

∫□d□ You can also use TI-Nspire CAS to graph this antiderivative using the indefinite integral template, also found in the Math template accessed by pressing ⟨⋈⟩. Figure 10-3 shows the results of using this alternative method.

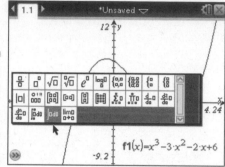
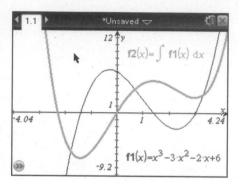

Figure 10-3:
Graphing
antideriva-
tives with
TI-Nspire
CAS.

To add a dynamic element, try inserting a slider (see Chapter 9) defined by
the variable c. As shown in the first screen in Figure 10-4, use the definite
integral template to graph the antiderivative as before. Then add $+ c$ in the
equation for the purpose of investigating the family of curves given by the
antiderivative of $y = x^3 - 3x^2 - 2x + 6$. Right-click the slider and change the settings to Minimized (see the second screen of Figure 10-4). As you see in the
last screen in Figure 10-4, you can click the slider and watch the graph of the
derivative translate vertically.

Figure 10-4:
Using a
slider to
investigate
the general
antide-
rivative of a
function.

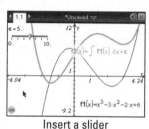
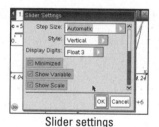
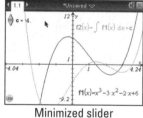

Insert a slider   Slider settings   Minimized slider

Part IV

The Geometry Application

In this part . . .

This part explores one of TI-Nspire's most dynamic applications. Using plane geometry, I show you how this application can be used to dynamically represent just about any geometric drawing or construction, all without a ruler or compass. I also show you how the Transformations tools can be used to create more advanced constructions. I then show you examples of how to work in both the plane geometry and analytic graphing windows simultaneously.

Chapter 11

Working with Geometric Objects

The Geometry application enables you to perform a variety of *Euclidean geometry* investigations. You can draw or construct triangles, circles, regular polygons, and parallel and perpendicular lines, as well as a host of other complex geometric objects. Whatever you can conjure up can be created in electronic form on TI-Nspire.

I cannot, of course, take you through every possible investigation that the Geometry application has to offer. Rather, it is my hope that you begin to gain an appreciation for what you can accomplish in this environment as well as the confidence to continue your own explorations.

Working in a Geometric Environment

In the following sections, I showcase those tools found in the Points & Lines menu and the Shapes menu.

Take advantage of the tooltips that are built in to the TI-Nspire. Just activate a tool and hover over the tool icon in the upper-left corner of the screen to make the tooltip appear. Even if you have used TI-Nspire for a while, you will probably discover something new when you read a tooltip. Maybe you already knew that to draw a line, you need to click two points (see Figure 11-1), but did you know that if you are in a coordinate plane environment (the Graphs application or using the analytic window in the Geometry application), you can just type the ordered pairs through which you would like the line to be drawn? Or, that you can press the Shift key to draw a horizontal line?

The Points & Lines Menu

In this section, I introduce you to various tools as I lead you through an investigation. My goal is to build a linear pair of angles that can be changed in a dynamic way, all the while proving that the angles are supplementary.

To begin the investigation, begin a new document and add a Geometry page (press 🏠on⇨New Document⇨Add Geometry).

The Line, Segment, Vector, and Ray tools are all drawn using nearly identical methods. Here are the steps for drawing a line and a ray:

1. **Press menu⇨Points & Lines⇨Line to access the Line tool.**

   See the second screen in Figure 11-1.

2. **Move your cursor anywhere on the screen and press 🔲 to mark a point through which the line will pass.**

   As you move the cursor, notice the ∅ symbol and the prompt "point." You can also use an existing point as the starting point for a line, segment, vector, or ray. As you move closer to an existing point, look for the ♘ symbol, which is TI-Nspire's way of telling you that you are locked in on that point.

3. **Move the cursor to another area on the screen and perform one of the following tasks:**

   • Press 🔲 to draw the line so that the line contains only the single point constructed during Step 2.

   • Press and hold ⇧shift, and then press 🔲 to draw a horizontal line (you can adjust the angle of the line in 15-degree increments).

   • Press tab followed by 🔲 to draw the line with a second point located at the current cursor location.

   • Move the cursor to an existing point. Look for the ♘ symbol, indicating that you are locked in on the point, and press 🔲 to draw the line.

4. **Press menu⇨Points & Lines⇨Ray to open the Ray tool.**

5. **Click once, then move the cursor to another area on the screen and press tab then 🔲 to draw the ray.**

The third screen in Figure 11-1 shows the Line and Ray tools in action.

  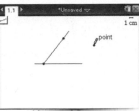

Figure 11-1:
Using
tooltips on
the Points &
Lines menu.

| Using tooltips | Points & Lines menu | Line and Ray tools |

As you know, lines go on forever in both directions, and rays go on forever in one direction. However, TI-Nspire typically displays lines and rays with a finite length. As shown in the first screen of Figure 11-2, move your cursor to the end of a line or ray (in the case of the ray, the end that *should* go on forever). When a blinking arrowhead appears, press ⌃ 🔲 to grab the end of the line (or ray) and use the Touchpad to extend or shorten its displayed length.

Note that you are changing the *appearance* of a line (or ray). These objects still behave as if they go on forever.

Using the Measurement Tool

Press menu⇨Measurement to access the Measurement menu. In Chapter 9, I talk about the Slope tool. In this chapter, I feature the Length, Area, and Angle tools.

Measuring angles

Remember, I am trying to create a linear pair of angles for this investigation. To measure one of these angles, follow these steps:

1. **Press menu⇨Measurement⇨Angle to open the Angle measurement tool.**

2. **Click the three points that form your angle, making sure that the second point you select is the vertex of the angle to be measured.**

   If you only have one point on your line, just press 🔲 somewhere else on the line to create a third point for your angle. See the second screen in Figure 11-2.

3. **Press esc to exit the Angle measurement tool.**

   Notice that the measurement changes dynamically when you grab the point on the ray and move the angle.

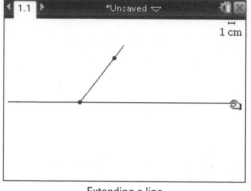
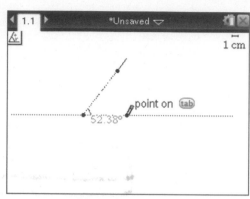

Extending a line

Measuring an angle

Figure 11-2:
Extending
lines and
measuring
angles.

Hiding the scale and changing the settings

Because this investigation does not require length measurement, you might also want to get rid of the scale in the upper-right corner of the screen. To do this, move your cursor to the scale, right-click ([ctrl] [menu]), and choose Hide Scale. See the first screen in Figure 11-3.

The default setting for a Geometry page measures angles in degrees. To display this measurement in radians, press [on]⇨Settings⇨Settings⇨Graphs & Geometry and change the Geometry Angle field of the dialog box from Degree to Radian. See the second screen in Figure 11-3. You can also access the settings by clicking the battery icon in the top-right corner of the screen. Because this angle was not created on a coordinate plane, it is considered a geometry angle. Angles that are created on a coordinate plane are considered graphing angles, for which the default setting is radians. Of course, for this investigation, I would prefer that you keep the Geometry Angle setting to be measured in degrees.

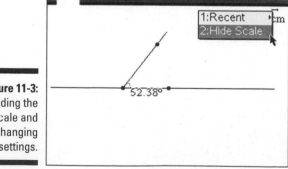
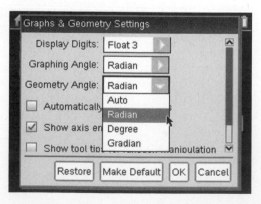

Figure 11-3:
Hiding the
scale and
changing
the settings.

The Shapes menu

Within the Shapes menu, you find the Circle, Triangle, Rectangle, Polygon, and Regular Polygon tools. Here are the steps to use the Circle tool:

1. **Choose** menu⇨**Shapes and select Circle from the list.**

   Hover over the Circle Icon in the upper-left corner of the screen. See the first screen in Figure 11-4.

2. **Move the cursor to the vertex of the angle and press** 🖐 **to mark the starting point.**

   For the Triangle, Rectangle, and Polygon tools, the starting point is a vertex; for the Circle and Regular Polygon, the starting point is the center.

3. **Use the Touchpad to adjust the size of the circle and press** 🖐 **to lock it in place.**

   As with the Line and Ray tools, you have the option of pressing tab and then 🖐 to draw a second point on the circle. Or, just click somewhere on the line and a point will automatically appear. See the second screen in Figure 11-4.

4. **Overlay a semicircle on the bottom half of the circle you just drew. Press** menu⇨**Points & Lines⇨Circle arc. Press** 🖐 **on one of the intersection points of the line and the circle, press** 🖐 **somewhere on the bottom half of the circle, and then press** 🖐 **on the other intersection point of the line and the circle.**

5. **Hide the circle. Move your cursor over the top half of the circle, right-click (** ctrl menu **), and choose Hide.**

   See the third screen in Figure 11-4.

Press esc to exit the current tool.

Figure 11-4:
Using the Circle and Circle arc tools.

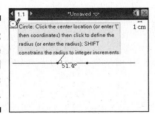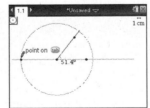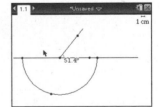

Exploring Constructions Dynamically

Remember, one of TI-Nspire's most powerful features is its ability to grab and move objects, enabling you to observe changes happening dynamically in real time. This is especially true of constructions, including those containing measurements and calculations.

Did you notice what happens when you drag the point on the ray? Try moving it both above and below the line. The angle changes, but it never gets larger than 180 degrees.

TI-Nspire can only measure angles that are 180 degrees or less (without help from advanced authors). The next tool that I am going to reveal to you is very helpful, especially if you begin making a geometric construction without thinking through each step in advance.

Using the Redefine tool to put the moveable point on a circle arc has some nice benefits. Placing the point on the circle arc will limit where I can drag the point; the point will follow the smooth path of the circle arc as it is dragged. Taking the time to perform this behind-the-scenes work will make your constructions look and feel more professional. Here are the steps to redefine a point:

1. **Press menu⇨Actions⇨Redefine. Click (🖫) the point on the ray, and then click (🖫) the circle arc.**

2. **Hover over the circle arc and right-click, ctrl menu⇨Hide. Repeat this step to hide the point you used to define the arc.**

   See the first screen in Figure 11-5.

Performing calculations and using the Attach feature

Calculations can be used to make sense of the measurements that were taken. The Attach feature is a handy way to display the results of the calculations and measurements. In your investigation, you use these tools to prove that two angles that form a linear pair are supplementary. Follow these steps to perform a calculation:

1. **Place some text for your calculation. Move your cursor to an open space, right-click (ctrl menu), and choose Text. Type a+b and then press enter.**

2. **Right-click (⌃ctrl ⌽menu) the text, choose Calculate, click (🖰) one of the angle measures, and then click (🖰) the other angle measure. Press ⏎enter to anchor the calculation.**

See the second screen in Figure 11-5.

3. **To lay down more text, right-click, ⌃ctrl ⌽menu ⇨Text. Press ?!▸ ⇨°(degree symbol). Press ⏎enter to anchor the text. Repeat two more times to give + and = a text box of their own.**

4. **Work right to left and attach each piece of the equation to the piece before it. Right-click (⌃ctrl ⌽menu) °(degree symbol), choose Attach, and click (🖰) the right side of the 180. Now, right-click 180 degrees and attach it to the = sign. Keep working backward until the equation is together in one attached piece!**

See the third screen in Figure 11-5.

Congrats! You have finished the first investigation in this chapter.

The Graphs and the Geometry applications are approximate environments. As a result, you may occasionally get a calculation or result that is not what you expect. For example, if you move the point to form a 0° angle, TI-Nspire displays 1.21E-6°, instead of 0°. This is scientific notation for the number 0.00000121, which is extremely close to zero. In most cases, you should consider these small noise values to be zero.

Figure 11-5:
Using the
Redefine,
Calculate,
and Attach
tools.

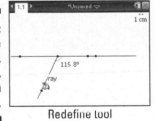

Redefine tool

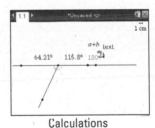

Calculations

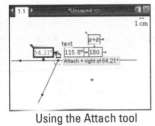

Using the Attach tool

Measuring length

With the Length Measurement tool, you can measure the perimeter of a polygon, the circumference of a circle, the length of a segment, or the distance between any two points. With that in mind, I would like you to construct an equilateral triangle ABC with side lengths of 10 inches.

First, draw a horizontal segment on a new Geometry page. Press `ctrl` `doc▾`⇨Add Geometry and then press `menu`⇨Points & Lines⇨Segment to access the Segment tool. Follow these steps to measure the length of the segment:

1. **Press `menu`⇨Measurement⇨Length to open the Length Measurement tool.**

2. **Move your cursor to the object being measured and press `⬚`.**

   To find the distance between two points, click each endpoint once to get the measurement to pop up.

3. **Use the Touchpad to move the measurement to a desired location and press `⬚` or press `enter` to drop it in place.**

4. **Press `esc` to exit the Length Measurement tool.**

   See the first screen in Figure 11-6.

Customizing the scale

Up to this point, I haven't said much about the scale (except how to hide it). By default, the scale is given in centimeters and is located in the upper-right corner of the screen.

Have you noticed that you have many clickable areas on the TI-Nspire? The scale behaves much like a text box. To change units, click the scale once and press the `del` key to delete the current unit. Type a new unit using the alpha keys, and press `enter` to complete the task. All existing length and area measurements are automatically updated to reflect this change. For this task, the sides of the equilateral triangle are measured in inches.

The second screen in Figure 11-6 shows the result of changing the scale from centimeters to inches.

The length of the segment that you measured is also a clickable area. Try double-clicking the length measurement and changing it to 10 in. See the third screen in Figure 11-6.

Figure 11-6: Working with length measurement and the scale.

Measuring a segment

Changing the scale

Adjusting the length

Using the Rotation tool

The Transformation menu provides five different transformations: Symmetry, Reflection, Translation, Rotation, and Dilation. The Rotation tool is a good choice for creating a triangle with specific angle requirements.

The angle of rotation can be defined by three points that lie on an angle or by a number.

To create an equilateral triangle, use the Rotation tool. I chose to use the numeric method to define my angle of rotation, which provides the advantage of being precise but with the drawback that it's a bit more static (although I can certainly edit my numerical value).

To perform a rotation, follow these steps:

1. **Press** menu⇨**Transformation**⇨**Rotation to open the Rotation tool.**

2. **Click the segment.**

3. **Click the point about which you want to rotate (I recommend the leftmost point on your segment).**

4. **Type the number (60) to represent the angle of rotation, and then press** enter.

5. **Press** esc **to exit the Rotation tool.**

The first screen in Figure 11-7 shows the result of using the Rotation tool. Notice that a text box with the angle of rotation (60) is still on the screen.

Positive rotations are in the counterclockwise direction, and negative rotations are in the clockwise direction.

Labeling objects

TI-Nspire allows you to add labels to any object. Use the Segment tool to finish drawing the last side of your equilateral triangle before you label the vertices. Use these steps to label the vertices of the triangle, A, B, and C:

1. **Press** menu⇨**Actions**⇨**Text to open the Text tool.**

2. **Move the cursor to the object (a point, line, rectangle, and so on) until the object starts blinking and its name appears. Press** 🖉 **to open a text box.**

Sometimes multiple objects are layered at the current cursor location (as indicated by the name of the current active object and the image of the tab key). Press tab to cycle through and activate each available object. Press 🖉 when the desired object's name appears to open the text box for that object.

3. **Type your label and press ⏎ to close the text box.**

   Repeat to label the other two vertices. Or, right-click (⌃ctrl ⌘menu) the points and choose Label. See the second screen in Figure 11-7.

4. **Press ⎋esc to exit the Text tool.**

When you manipulate a drawing or construction, object labels stay with the object. An object label can only be moved within a small distance of the object with which it is associated.

You have just constructed an equilateral triangle! Before you go on to the next investigation, I want to point out a few things about the triangle you have just built that are in sharp contrast with the one built using the Shapes menu. Whenever you use the Shapes menu, this time-saving procedure can be used to label the vertices. To construct and label a second triangle with vertices D, E, and F, follow these steps:

1. **Press ⎯menu⇨Shapes⇨Triangle to initiate the Triangle tool.**

2. **Move your cursor to an open space and press 🖘 to place the first vertex of the triangle.**

3. **Use the alpha keys to label the vertex. For the first vertex, press ⇧shift D.**

4. **Repeat Steps 2 and 3 to place and label the other two vertices.**

   See the third screen in Figure 11-7.

Figure 11-7:
Using the
Rotation tool
and labeling
objects.

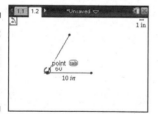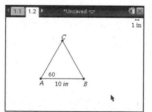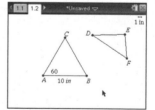

Moving and resizing objects

You can move or resize any object. However, you face some constraints based on the type of drawing you have constructed. For example, consider the equilateral triangle. This object can be grabbed and moved or resized, but because of the way that this triangle was built, you are limited in what can be grabbed. Hover over each side. Do you notice that only one side changes thickness? You will be able to grab and move that side only.

Similarly, when you hover over the top of the vertices, only one point gets larger: the point that you can grab and move.

The triangle remains an equilateral triangle unless you change the angle of rotation. Try double-clicking the text box containing 60 and changing it to 90. Do you recognize what type of special triangle it has become? See the answer in the first screen in Figure 11-8.

To move or resize an object, you must position the cursor on the object (or a part of the object) and press [ctrl][⇧] to grab the object (as indicated by the ✍ symbol). Then use the Touchpad to move the object. To translate the entire triangle, grab any side. To change the shape of the triangle, grab one of the vertices.

A circle moves and resizes in surprising ways. Use the Shapes menu to activate the Circle tool. Press [menu]➪Shapes➪Circle. Use the Circle tool to draw a circle on your screen. See the second screen in Figure 11-8.

If you are unfamiliar with how to operate a particular tool, take advantage of the tooltips. Simply hover over the tool icon in the upper-left corner of the screen and read the tooltip.

Suppose that you want to change the size of your circle. You may first try grabbing the center point, only to find out that this action translates the circle. Grab the circle itself to change the circle's size.

As you move near an object, TI-Nspire displays the word that describes the current object that can be grabbed. If multiple objects are near the cursor, TI-Nspire also displays the [tab] symbol (see the third screen in Figure 11-8). By pressing [tab] repeatedly, you can cycle through all the objects near the cursor location until you find the one that you want to grab (or label, hide, change attributes of, and so on).

Figure 11-8:
Moving and
resizing
objects.

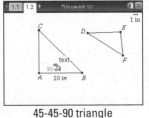

45-45-90 triangle

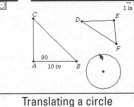

Translating a circle

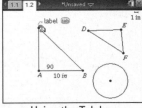

Using the Tab key

Selecting multiple objects

You may want to select multiple objects for two reasons. First, multiple selected objects can be translated. Second, all selected objects can be deleted by pressing the [del] key once.

To move all objects on a screen, move your cursor to open space and press [ctrl][⊡]. You see the crunched-up paper symbol (✥), indicating that all objects on the screen can be translated using the Touchpad. See the first screen in Figure 11-9.

Here are two methods that can be used to select multiple objects:

> ✔ **Method 1:**
>
> 1. Move to an object and press [⊡]; it will begin to flash.
>
> 2. Continue moving to additional objects, pressing [⊡] each time until they begin to flash.
>
>   To deselect individual objects from a group, move the cursor to the desired object and press [⊡]. To deselect all objects, press [esc].
>
> ✔ **Method 2:**
>
> 1. Press [menu]⇨Actions⇨Select to invoke the Select tool.
>
> 2. Consider a rectangle surrounding the objects that you want to select. Move to one of the vertices of this rectangle and press [enter].
>
> 3. Move to the opposite corner of this rectangle and press [enter] again to select the group of objects.

Any object that is even partially contained within the rectangle is selected. See the second screen in Figure 11-9.

After objects have been selected, press the [del] key to delete them.

To move selected objects, simply use the Touchpad keys. Alternatively, move the cursor to one of the selected objects, grab the object ([ctrl][⊡]), and use the Touchpad to move the group. See the third screen in Figure 11-9.

Figure 11-9:
Selecting
multiple
objects.

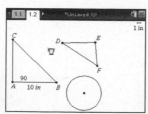 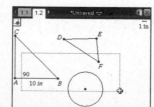 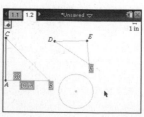

If you want to delete all objects in an application, press [menu]⇨Actions⇨Delete all. Press [enter] at the warning prompt to delete the objects, or press [esc] if you decide against it.

Changing an Object's Attributes

In Chapter 9, I talk about how to change the attributes of a graph. You can also change the attributes of a geometry object. Press [menu]⇨Actions⇨Attributes to invoke the Attributes command. Then, move to the object whose attributes you want to change and press [📷] or [enter]. Alternatively, move your cursor to the object whose attributes you want to change and press [ctrl][menu]⇨Attributes.

After opening the Attributes menu, press the ▲▼ keys to move through the different options (weight, style, and so on). Use the ◀▶ keys to view the different options within each attribute and notice the corresponding attribute change on the graph as well.

To put an attribute into effect, you must use the Touchpad keys to select it and then press [enter]. To exit the Attributes menu without making a change, press [esc]. See the first screen in Figure 11-10.

The Color menu can be accessed from the context menu for an object. Move your cursor over the perimeter of the circle you created, right-click ([ctrl][menu]), choose Color, and select Fill Color. Choose your favorite color from the 16 choices. This tool can be used to change the interior shading of any closed figure drawn using the Shapes menu. So, unfortunately, you will not be able to use Fill Color in triangle ABC (unless you use the Triangle tool from the Shapes menu to draw a triangle on top of triangle ABC). See the second screen in Figure 11-10.

You can use the Fill Color tool with multiple shapes at one time. Just click ([📷]) the shapes that you want and then right-click ([ctrl][menu]) one of them and choose Color. See the third screen in Figure 11-10.

Changing the line color of an object will also cause the label of the object to change color to match the line color.

Figure 11-10:
Changing an object's attributes and fill color for shapes.

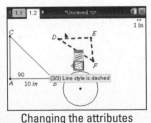
Changing the attributes

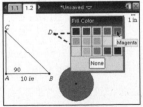
Using the Fill Color tool

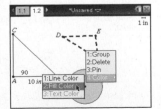
Selecting multiple shapes

Measuring Area

The Area Measurement tool allows you to measure the area of a circle, polygon, or anything else that was constructed using the Shapes menu. Here are the steps to follow:

1. **Press** [menu]⇨**Measurement**⇨**Area to open the Area Measurement tool.**

2. **Move your cursor to the object being measured and press** [✥].

3. **Use the Touchpad to move the measurement to a desired location and press** [✥], **or press** [enter] **to drop it in place.**

4. **Press** [esc] **to exit the Area Measurement tool.**

Did you try to measure the area of triangle ABC? How did that work for you? Because I didn't use the Shapes menu to create it, the Area Measurement tool simply does not work. If you needed to measure the area of the triangle, you could use the Triangle tool to draw a triangle on top of the segments.

Using the Lock and Pin Features

For this investigation, insert a new Geometry page (press [ctrl][I] and choose Add Geometry). Now draw a triangle with a horizontal base (press [menu]⇨Shapes⇨Triangle; don't forget to hold the [⇧shift] key down to construct a horizontal line).Then follow these steps to construct an ellipse:

1. **Secure the two vertices of the triangle for the side that is horizontal, which will become the foci of the ellipse. Move your cursor over the vertex, right-click (**[ctrl][menu]**), and choose Pin.**

   See the first screen in Figure 11-11.

2. **Repeat Step 1 for the second vertex.**

3. **Measure the perimeter of the triangle. Press** [menu]⇨**Measurement**⇨**Length. Move your cursor over the triangle until** ✥ **appears, then click (**[✥]**) twice to anchor the measurement.**

When finding the perimeter of a polygon, notice that the word describing the polygon (triangle, rectangle, or polygon) appears along with the [tab] symbol. Press [tab] to switch to finding the length of the side of the polygon corresponding to the current cursor location. Press [tab] again to switch back to finding the perimeter of the polygon.

4. **Move your cursor over the perimeter measurement, right-click,** [ctrl] [menu]⇨**Attributes, and use your Touchpad to lock the measurement.**

   See the second screen in Figure 11-11.

5. **Press** menu⇨**Trace**⇨**Geometry Trace. Move your cursor over the third vertex of the triangle, click the vertex to indicate that you want the geometry trace of that point, grab the point (** ctrl 🔳 **), and then use the Touchpad to move it around to form the ellipse.**

If a point is pinned, it can't be moved unless it is unpinned. Other geometric objects and text can also be pinned, which is a nice feature, because it prevents someone from accidentally moving the wrong point or grabbing and moving a label instead of a point.

The third screen in Figure 11-11 shows a nifty mathematical reason for locking the perimeter of a triangle. By using Geometry Trace, you can trace the path of the vertex of a triangle with a fixed perimeter. The tracing results in an ellipse because the sum of the lengths of the two sides of the triangle that meet this vertex must be constant. How cool is that?

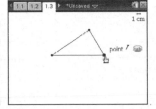 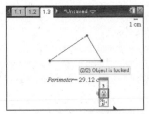 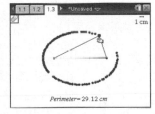

Figure 11-11:
Using the Lock and Pin features.

You may also want to try locking the perimeter of a rectangle, which can facilitate finding the maximum area for a fixed perimeter.

Constructing Geometric Objects

The act of creating a geometric construction can provide a rich mathematical experience, especially when used with the dynamic click-and-drag feature of TI-Nspire. To access the Construction menu, press menu⇨Construction.

Perpendicular lines

To construct a perpendicular line, you must already have a segment, line, or ray. Then follow these steps:

1. **Press** menu⇨**Construction**⇨**Perpendicular to open the Perpendicular tool (you may want to hover over the icon in the upper-left corner of the screen).**

2. **Click the segment, line, or ray through which you want your perpendicular line to pass.**

3. **Move the cursor to an existing point or to any point on the screen through which you want the perpendicular line to pass.**

As you move the cursor, notice that a ghosted image of the perpendicular line is displayed on the screen.

4. **Press 🔲 to set the perpendicular line in place.**

5. **Press 🔲 to exit the Perpendicular tool.**

Steps 2 and 3 can be reversed.

Perpendicular Bisector tool

To use the Perpendicular Bisector tool, you must have a segment or an *implied segment* (two points). In this investigation, you are going to use the Perpendicular Bisector tool to construct a parabola. Insert a new Geometry page and follow these steps to construct the sketch:

1. **Draw a horizontal line that will represent the directrix of the parabola. Press menu⇨Points & Lines⇨Line. Click (🔲), press ⬆shift, maneuver the line, and press 🔲 to place the line.**

2. **Place a point above the line that represents the focus of the parabola. Press menu⇨Points & Lines⇨Point. Click (🔲) to place the point. Repeat this step to place a point on the directrix line.**

See the first screen in Figure 11-12.

3. **Activate the Perpendicular Bisector tool: Press menu⇨Construction⇨Perpendicular Bisector. Click (🔲) the two points that you just created, which are endpoints of the *implied segment* through which you want your perpendicular bisector to pass.**

Your parabola will look better if you grab the ends of the perpendicular bisector and extend the line. See the second screen in Figure 11-12.

4. **Use the Locus tool to create your parabola. Press menu⇨Construction⇨Locus. Click (🔲) the perpendicular bisector and then click (🔲) the point that is on the directrix.**

See the third screen in Figure 11-12.

5. **Press 🔲 to exit the Locus tool.**

Figure 11-12:
Using the
Perpen-
dicular
Bisector
tool to
create a
parabola.

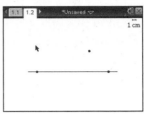 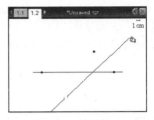 

Angle Bisector tool

To use the Angle Bisector tool, follow these steps:

1. **Press menu⇨Construction⇨Angle Bisector to open the Angle Bisector tool (don't forget about the icon located in the upper-left corner of the screen).**

2. **Click the three points that form your angle, making sure that the second point you select is the vertex of the angle to be bisected.**

3. **Press esc to exit the Angle Bisector tool.**

As with the other tools described in this section, the Angle Bisector tool creates a line rather than a segment or ray.

Although it's nice to "see" the angle that you are bisecting, all you need to use this tool are three points.

Parallel lines

The steps for constructing a parallel line are identical to those steps used to construct a perpendicular line. Just follow the steps in the earlier section, taking out the word *perpendicular* and inserting the word *parallel*.

3. **Use the Polygon tool to draw the trapezoid. Press ⎡menu⎤⇨Shapes⇨ Polygon. Click (⌧) the location that you want for each of the four vertices. Then click (⌧) the first vertex again to close the trapezoid.**

   Hide the parallel lines and you have it. See the second screen in Figure 11-13.

4. **Press ⎡esc⎤ to exit the Parallel tool.**

> TIP
>
> Sometimes you want to construct perpendicular or parallel segments or rays, not lines. To accomplish this task, construct the perpendicular or parallel line. Then open the Segment (or Ray) tool as you normally do. As you move the cursor to the perpendicular or parallel line, the words `Point On` appear, telling you that the segment will be associated with the line. Watch for `Point On` again as you locate the endpoint of your segment (or specify the direction of the ray). After you draw the segment or ray, use the Hide/Show tool to hide the line, and you are all set.

Midpoints

To use the Midpoint tool, you must have a segment or an *implied segment* (two points). Here is a math question for you: What shape do you get when you connect all four midpoints in a quadrilateral? To find out the answer, try the construction on the trapezoid that you just created. Here are the steps to construct the shape:

1. **Press ⎡menu⎤⇨Construction⇨Midpoint to open the Midpoint tool (notice the icon located in the upper-left corner of the screen).**

2. **Click the segment to construct the midpoint, or click the endpoints of the implied segment. Mark the midpoint for all four sides of the trapezoid.**

3. **Use the Polygon tool to connect the midpoints. Press ⎡menu⎤⇨Shapes⇨ Polygon. Click (⌧) each of the four midpoints, and then click (⌧) the first midpoint again to close the polygon.**

   See the third screen in Figure 11-13.

4. **Press ⎡esc⎤ to exit the Polygon tool.**

Figure 11-13: Using the Parallel and Midpoint tools.

 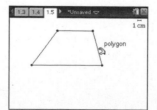 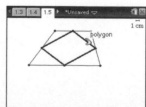

Investigating Transformations

The Transformation menu provides five different transformations: Symmetry, Reflection, Translation, Rotation, and Dilation. Earlier in the chapter, I used the Rotation tool to create an equilateral triangle. Here's a brief description of what each of the other four transformation tools can do and what you do to perform the transformation:

- ✔ **Symmetry:** This transformation gives the image of an object with a 180-degree rotation about a point.

- ✔ **Reflection:** This transformation gives the image of an object reflected over a line or segment.

- ✔ **Translation:** This transformation gives the image of an object translated by a distance and direction given by a vector or two points.

- ✔ **Dilation:** This transformation gives the image of an object with a point that is the center of dilation and a number specifying the dilation factor.

The first two transformations, symmetry and reflection, are accomplished using similar steps. Here are the steps to reflect a polygon over a line:

1. **Draw a polygon.**

2. **Draw a segment or line.**

3. **Press** menu⇨**Transformation**⇨**Reflection to open the Reflection tool.**

4. **Click the polygon.**

5. **Click the segment or line.**

    When using the Symmetry tool, click a point.

6. **Press** esc **to exit the Reflection tool.**

Here are the steps for using the Translation tool:

1. **Draw an object.**

2. **Draw a vector (or two points) that gives the direction and length of the translation.**

3. **Press** menu⇨**Transformation**⇨**Translation to open the Translation tool.**

4. **Click the object.**

5. **Click the vector (or each of the two points).**

6. **Press** esc **to exit the Translation tool.**

The first screen in Figure 11-14 shows the results of using the Symmetry, Reflection, and Translation tools. I started by drawing polygon *A*. I then used the Symmetry tool to complete a 180-degree rotation of polygon *A* about point *p*, which gave me polygon *B*.

I then used the Translation tool to translate polygon *A* to polygon *C*. To complete this step, I clicked polygon *A* and then the translation vector located at the bottom of the screen (to draw a vector, press menu⇨Points & Lines⇨Vector). Notice that the corresponding points on polygons *A* and *C* are located at a distance and direction equal to the translation vector.

Finally, I used the Reflection tool to complete the reflection of polygon *C* over the line. Try dragging objects and observing the corresponding changes.

To perform a dilation, follow these steps:

1. **Draw a regular pentagon with the Regular Polygon tool. Press menu⇨Shapes⇨Regular Polygon to open the tool.**

2. **Move the cursor to an existing point or any open space on the page and press ⌖ to mark the starting point.**

3. **Use the Touchpad to move the cursor away from the center of the regular polygon and press ⌖ to establish the radius and first vertex of the regular polygon.**

4. **Use the Touchpad to move the cursor in a clockwise direction. This action decreases the number of sides of the regular polygon. Press enter when the number of sides is 5.**

   Take a minute to play around with this tool. You can have up to 16 sides, and you can create stars and other shapes when you move to less than three sides. See the second screen in Figure 11-14.

   In the classroom, you can use the Regular Polygon tool as a good visual to dynamically demonstrate that the more sides that you have, the closer the area of the polygon gets to the area of a circle with the same radius.

5. **Press menu⇨Transformation⇨Dilation to open the Dilation tool.**

6. **Click the point that defines the center of dilation.**

7. **Type a number to give the scale or dilation factor; then press enter.**

8. **Click the object, which in this case is the pentagon.**

9. **Press esc to exit the Dilation tool.**

The third screen in Figure 11-14 shows the result of using the Dilation tool. On the left, I show a dilation that results in an enlargement with a scale factor of 2 and center of dilation at point *P*. The image is shown with lines of medium weight, and the preimage is shown with lines of thin weight.

On the right, I show a dilation that results in a reduction with a scale factor of 2/3.

Figure 11-14: Using the Transformation menu.

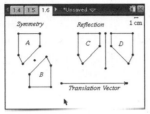 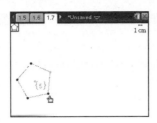 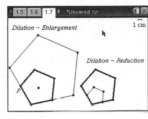

Animating Objects

You can animate a point (or multiple points) on a line, ray, axis, vector, graph, segment, or circle. Assume that you want to animate point *a* on a circle. As you do so, point *a* rotates in a counterclockwise direction about the circle and, consequently, the polygon rotates about point *p*.

Here's the easiest way to perform this animation:

1. **Move the cursor to point *a* and press** ctrl menu⇨**Attributes.**

2. **Press ▼ to select the animation attribute.**

The words Unidirectional animation speed indicate that the point will move in only one direction, counterclockwise. Unidirectional animation speed is the only option associated with this attribute when animating a point on a circle, line, or ray. However, I have discovered that if you enter a negative value for the speed, it will move in the clockwise direction! How cool is that?

If you animate a point on a segment, you have the option of specifying "Alternating animation speed." In this scenario, the point moves to one endpoint, switches direction, and moves back to the other endpoint.

3. **Type a number from 1 (slow) to 12 (fast) and press Enter.**

The animation automatically begins.

4. **Press ◀▶ to change whether the animation is unidirectional or alternating direction.**

This feature is not available when you animate a point on a circle.

5. **Press** enter **to close the Attributes menu and open the animation control panel.**

The first screen in Figure 11-15 shows the animation attribute. The second screen shows the animation control panel. The object can be moved to any location on the screen. The left button on the animation control panel is used to reset the animation. Move the cursor over this button, press 🔲 to stop the animation, and return the animated point to its original location.

The right button toggles between *pause animation* (second screen) and *start animation* (third screen).

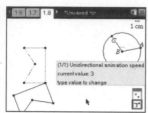 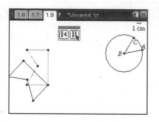 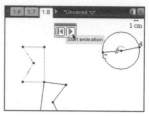

Figure 11-15:
Animating
points and
rotation by
three points.

To stop an animation and remove the control panel, follow these steps:

1. **Stop the animation using the animation control panel.**

2. **Move the cursor back to the animated point and press `ctrl` `menu`⇨Attributes.**

3. **Press ▼ to select the animation attribute.**

4. **Type 0 and `enter`.**

Using Conditional Statements

Conditional attributes allow you to dynamically hide, show, or change the color of text or any geometry object or graph. Follow these steps to create an interactive document using conditional statements:

1. **Press `menu`⇨Points & Lines⇨Ray and create an angle using two rays.**

2. **Press `menu`⇨Measurement⇨Angle and measure the angle.** Make sure the geometry angle is set to degrees by pressing `menu`⇨Settings.

3. **Hover your cursor over the angle measurement, press `ctrl` `menu`, and choose Store.** Enter **theta** and then press `enter` to store the angle measurement.

4. **Press `ctrl` `menu` in an open space, choose text, and enter obtuse.**

5. **Hover your cursor over the obtuse text box, press `ctrl` `menu`, and choose Conditions.**

6. **Enter theta>90 in the Show When dialog box.** The text box appears only when the angle measurement is greater than 90 degrees.

7. **Enter 7 in the Line Color dialog box.** When the text appears, it will be red. There are 15 different colors indicated by the numbers 1–15; click the Colors button to view the entire color pallete.

8. **Hover your cursor over the angle measurement, press `ctrl` `menu`, and choose Conditions.**

9. **Enter 4 in the Line Color dialog box.** Because nothing is in the Show When dialog box, the angle measurement displays in green text.

Chapter 12

Using an Analytic Window in the Geometry Application

In This Chapter

▶ Working with the analytic window in the Geometry application

▶ Using the analytic view to discover a geometric formula

▶ Taking a look at some other examples

*Y*ou have many reasons for working strictly in the Graphs application or the Geometry application. Sometimes, however, it's to your advantage to work in an environment that contains the best of both worlds. In this chapter, I show you how to set up such an arrangement. I then provide you with some examples that give compelling reasons for why using the analytic window is such a good option.

Adding an Analytic Window to a Geometry Application

Here are the steps to follow to set up a the Geometry application with an analytic window:

1. **Open a new Geometry page. Press** [ctrl] [doc▾]⇨**Add Geometry.**

2. **Press** [menu]⇨**View**⇨**Show Analytic Window.**

The first screen in Figure 12-1 shows that these steps place a small analytic window near the lower-left corner of the screen. In the second screen, I show that the boundaries of the analytic window can be expanded by grabbing the end of an axis. Sometimes, it is difficult to grab the end of the axis. Press [tab] while hovering over the end of the axis to cycle through the layers of objects. Press [ctrl] [✥] to grab when the word *axes* appears. Use the Touchpad keys to adjust the boundaries of the analytic window. See the third screen in Figure 12-1.

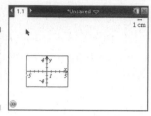 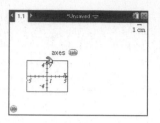 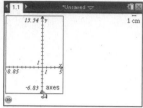

Figure 12-1:
Setting up
an analytic
window.

Exploring the Differences between Analytic and Geometric Objects

Figure 12-2 does a nice job of showing how objects behave differently in the analytic window. You see two circles that have the same radius, but look quite different. The circle on the left was drawn in the analytic window, whereas the circle on the right was drawn in the Geometry work area.

Here are the steps to draw a circle with radius 4 outside the analytic window (in the Geometry work area):

1. **Use the Text tool and press ④ to represent the length of the radius of the circle.**

2. **Press menu⇨Shapes⇨Circle, press 🖲 to choose the location of the center, click the 4, and then click again to anchor the circle.**

Caution! You will get an error message, `Numeric entry not available in mixed views,` if you try to type a number for the radius using the Circle tool. The following steps show another way to draw a circle with radius 4:

1. **Press menu⇨View⇨Grid⇨Dot Grid to reveal a grid in the analytic window.**

   The grid corresponds to the tick marks located on each axis. A grid enables you to draw circles with a radius of exactly 4 units.

2. **Press menu⇨Shapes⇨Circle, press 🖲 to choose the location of the center (the origin for this example), and then click the tick mark that corresponds to 4 units.**

   See the second screen in Figure 12-2. If you resize the circle, it will jump to other grid locations.

I pressed menu⇨Actions⇨Coordinates and Equations to find the equation of the analytic circle on the left (you can also right-click the circle and choose Coordinates and Equations). You can use this tool to find the equations of circles, lines, and tangents as long as they are drawn in the analytic window. If these same objects are drawn in the Geometry work area, the Coordinates and Equations option is not available.

Likewise, I've used the Measurement (length) tool to find the circumference of both circles. Notice that the circumference measurement for the circle on the left is "unitless" (as indicated by the letter u), whereas the units for the circle on the right are given in centimeters. Keep in mind that all measurements found in an analytic window are unitless; the scale displayed in the upper-right corner of the screen applies only to plane geometry objects.

Finally, in the third screen in Figure 12-2, I changed the scale for the y-axis in the analytic window, and the analytic objects changed accordingly. Keep in mind that two-dimensional objects will become distorted if you adjust the scale on only one axis. Did you notice that the circle on the right was unaffected by the change of scale in the analytic window? The aspect ratio is always 1:1 for plane geometry objects.

Figure 12-2: Observing the differences between analytic and plane geometry objects.

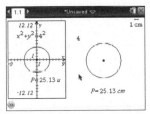 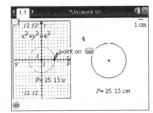 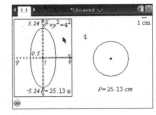

If you move a plane geometry object to the analytic window, it continues to behave as a plane geometry object instead of changing automatically to an analytic object.

The Power of Multiple Representations

Many geometric properties can be modeled algebraically. For example, the sum, S, of the interior angles of an n-sided convex polygon can be represented by the algebraic formula $S = 180(n - 2)$.

By working in an environment with both an analytic window and a plane geometry work area, TI-Nspire provides the opportunity to explore these geometric and algebraic relationships. Furthermore, the dynamic, click-and-drag technology inherent to TI-Nspire makes these relationships even more evident.

Discovering an area formula

I can use a mixed view with an analytic window open on a Geometry page to explore the formula for the area of a circle.

I've also decided to use the Automatic Data Capture feature to generate a scatter plot of the area-versus-radius data. This approach is a bit complicated. However, this option has the benefit of adding numeric data to a Lists & Spreadsheet page.

If you attempt this construction, the circle is a bit tricky. First draw a horizontal line. Overlay a segment on top of the line and hide the line. Draw your circle so that the segment is the radius of the circle.

Here's a brief description of how I accomplish this task (see Figure 12-3). For more information about configuring TI-Nspire for Automatic Data Capture, refer to Chapter 16.

1. **Click each measurement once and press var⇨Store Var. Type a descriptive name for each variable (such as *radius* and *area*). Notice that variable names appear in bold font. See the first and second screens in Figure 12-3.**

2. **Configure a Lists & Spreadsheet page for Automatic Data Capture.**

3. **Configure the entry line to draw a scatter plot of the area-versus-radius data.**

4. **Drag the point on the edge of the circle, and watch the scatter plot appear in the analytic window.**

   See the last screen in Figure 12-3.

Figure 12-3: Performing a data capture on area-versus-radius data.

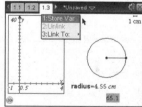 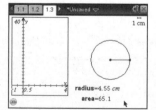 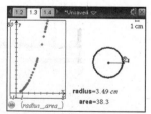

The scatter plot in this exploration is tied to the analytic window, which means if I adjust the window settings, the scatter plot adjusts accordingly.

It would appear that the scatter plot is quadratic — but what is the equation that models this data? Configure the entry line for function graphing and graph the function $y = x^2$.

As shown in the first screen in Figure 12-4, move the cursor to the graph until the ✕ appears and press [ctrl][🔲] to grab the edge of the graph. As you see in the second screen in Figure 12-4, use the Touchpad keys to manipulate the graph to pass through the scatter plot and watch the equation update accordingly. It seems pretty clear that the equation $y = \pi x^2$ (better known as $A = \pi r^2$) is the formula that relates the area of a circle to its radius.

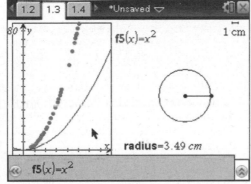 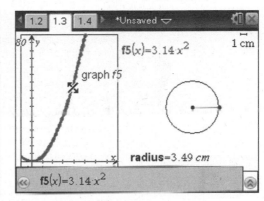

Figure 12-4: Finding the formula for the area of a circle.

Other examples

I think the two examples included in this chapter do a nice job of illustrating the benefits of using the Plane Geometry view and an analytic window together. Don't forget to use the Activities Exchange at education.ti.com to find some more ideas that utilize this same environment. The following list includes brief descriptions of a few more explorations that work quite nicely in the simultaneous environment of the analytic window and the Geometry work area:

✔ **Finding the maximum area of a rectangular fenced-in area that uses a barn as one side.**

This problem represents a variation of a problem I feature in Chapter 16. I use the Manual Data Capture feature to draw the scatter plot. See the first two screens in Figure 12-5 to see how the problem looks.

✔ **Finding the minimum surface area of a cylinder with fixed volume.**

This is another classic optimization problem in which a quantity must be minimized. To get really fancy, construct a model with a fixed volume of 355 milliliters (the volume of a standard soda can) and use Geometry Trace, the Locus tool, or Data Capture to form an analytic representation. Or, search for optimization in TI's Activities Exchange to find a previously made, and editable, version of this problem. See the third screen in Figure 12-5.

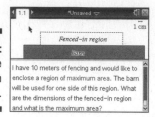
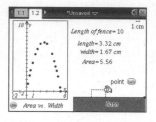
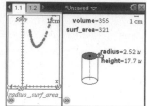

Figure 12-5:
A few more
optimization
problems.

✔ Finding the formulas for the area and volume of geometric objects.

As with the area-of-the-circle problem in this chapter, a geometric model of virtually any two- or three-dimensional object can be explored analytically.

In Figure 12-6, I constructed a model of a cone whose height and radius can be varied (by dragging on their respective points). The first two screens show the corresponding scatter plot of the volume versus radius for a constant height. The third screen shows the graph of the function that models this scatter plot.

Using similar methods, you can also explore the relationship between height and volume for a constant radius.

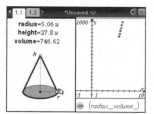
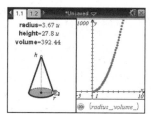
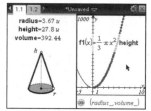

Figure 12-6:
Exploring
the volume
of a cone.

This construction contains a few tricks. For example, I drew the circle on a coordinate grid and adjusted the y-axis scale to squish the circle and make it look more like an ellipse. I then used the Segment tool to draw the sides of the cone that meet at the vertex (located on the y-axis). Finally, I hid the axes. Because I can only use one Analytic view per page, I split the page in two and added an Analytic view on the right, which is where I graphed the scatter plot using the Automatic Data Capture feature.

Part V
The Lists & Spreadsheet Application

The 5th Wave

By Rich Tennant

"Okay — let's play the statistical probabilities of this situation. There are 4 of us and 1 of him. Phillip will probably start screaming, Nora will probably faint, you'll probably yell at me for leaving the truck open, and there's a good probability I'll run like a weenie if he comes toward us."

In this part . . .

This part deals with the numeric side of mathematics. I cover the tools and methods that are commonly used with most generic computer spreadsheet applications. I then get into some of the more TI-Nspire–specific tasks related to this application such as constructing scatter plots, performing regressions, and using Data Capture. These tasks will convince you that the Lists & Spreadsheet application forms a great partnership with both the Graphs and Geometry applications.

Chapter 13

Applying What You Already Know about Spreadsheets

*I*f you have some familiarity with computer spreadsheets, you'll feel right at home with the Lists & Spreadsheet application. Regardless of your comfort level with spreadsheets, this chapter offers you an overview that is sure to orient you to the basic structure of the Lists & Spreadsheet application.

In Chapters 14 through 16, I address more advanced features of the Lists & Spreadsheet application, including those features that are unique to TI-Nspire.

Understanding Row, Column, and Cell References

 To open a new Lists & Spreadsheet page, press `ctrl` `doc▾` and select Add Lists & Spreadsheet from the list. Alternatively, press `⌂on` and select the Lists & Spreadsheet icon from the available options. Figure 13-1 shows a blank Lists & Spreadsheet page as well as a description of its various components.

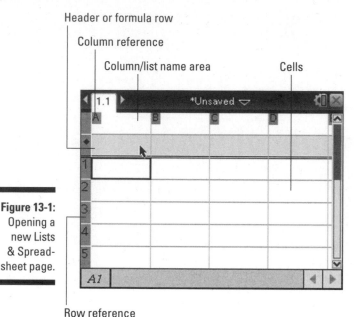

Header or formula row

Column reference

Column/list name area

Cells

Row reference

Figure 13-1:
Opening a
new Lists
& Spread-
sheet page.

Each Lists & Spreadsheet page contains a total of 26 columns. Each column is labeled with a letter, starting with A and ending with Z.

Each Lists & Spreadsheet page has a total of 2500 rows. Each row is labeled with a number, starting with 1 and ending with 2500.

In Figure 13-1, notice that a dark border surrounds the cell A1. This name is given to the cell because it is located in column A, row 1. When naming cells, make sure that you start with the column reference (a letter) followed by the row reference (a number). The cell reference for a highlighted cell is located in the lower-left corner of the screen. Sometimes you have a rectangular block of cells highlighted. In this case, you see the cell reference associated with the upper-left corner of the rectangle, followed by a colon, followed by the cell reference associated with the lower-right corner of the rectangle. See Figure 13-2.

Cell reference   Cell reference   Cell reference

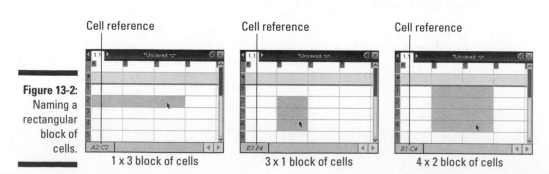

Figure 13-2:
Naming a
rectangular
block of
cells.

1 x 3 block of cells   3 x 1 block of cells   4 x 2 block of cells

TIP

To highlight a rectangular block of cells, move to a corner of the rectangular block, press and hold the ⌂shift key, use the Touchpad keys to move to the corner diagonally opposite your starting point, and then release the ⌂shift key.

Naming Columns

Each column already has a letter reference that can be used to perform mathematical computations on columns of data. To reference a column in a formula, simply type the letter of the column followed by a set of brackets (ctrl ( ). I show you an example of how to use this type of reference in Chapter 14.

Additionally, TI-Nspire offers you a way to name your columns with a word that helps to convey what the data represents. Consider, for example, that you want to analyze data on cellular telephone subscriptions for a range of years. Here are the steps to follow to give your data a descriptive name:

1. **Move the cursor to the column list name area.**

   This is the white box located at the top of a column.

2. **Type the name of your list using the alpha keys.**

   As you can see in the first screen in Figure 13-3, I used the word *year* as the list name for column A.

   List names follow the same rules as variable names. Refer to Chapter 6 for a description of how to name variables.

   REMEMBER

3. **Press tab to move the next column and type its name.**

   I've typed the name *subscriptions* to remind me that the data in the second list refers to the number of cellular telephone subscriptions (in millions) for each specified year. See the second and third screens in Figure 13-3.

4. **Continue to press tab to name additional columns. Press enter after you have named your last column of data.**

   Don't worry that you cannot see the full list name. I'll tell you how to widen these columns later in the chapter.

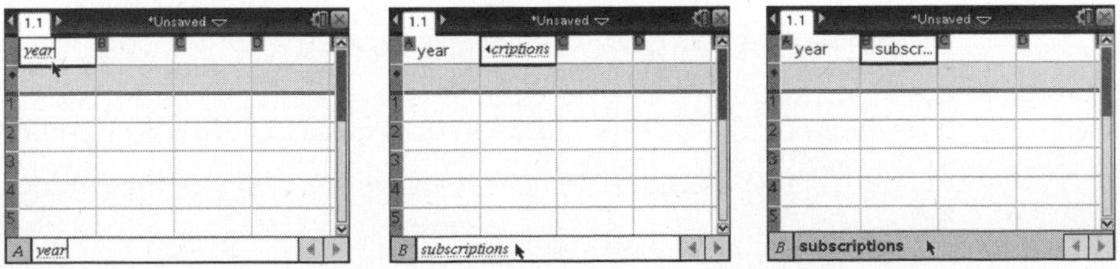

Figure 13-3:
Naming columns.

Moving around in Lists & Spreadsheet

Recall that a dark outline designates your current location in a spreadsheet. To enter data or a formula, simply start typing and press the `enter` key when you have finished. This moves you to the cell beneath the one where you are currently located.

Alternatively, press the `tab` key after entering data or a formula in a cell. This moves you to the cell immediately to the right of the cell where you are currently located.

In Figure 13-4, I have entered the data that represents the number of cell-phone subscriptions (in millions) for years since 1985.

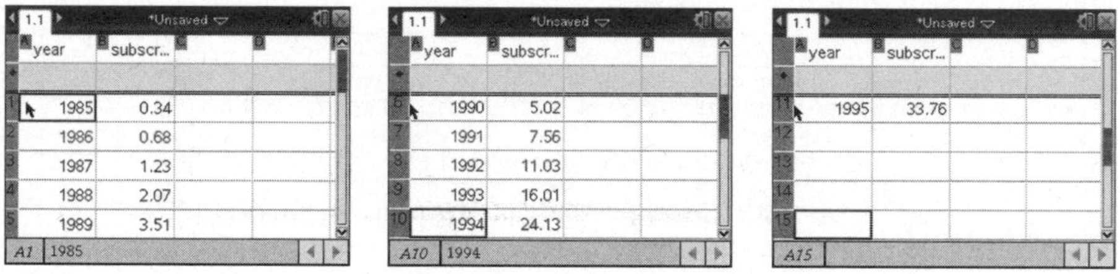

Figure 13-4:
Entering
data.

1.1	*Unsaved	
year	subscr...	
1985	0.34	
1986	0.68	
1987	1.23	
1988	2.07	
1989	3.51	
A1	1985	

1.1	*Unsaved	
year	subscr...	
1990	5.02	
1991	7.56	
1992	11.03	
1993	16.01	
1994	24.13	
A10	1994	

1.1	*Unsaved	
year	subscr...	
1995	33.76	
A15		

In general, you can move around the spreadsheet by pressing the arrow keys on the Touchpad. Here are some other tricks that can help you move around a spreadsheet more efficiently:

✔ **Press `menu`⇨Actions⇨Go To.**

After you select this command, a dialog box appears. Simply type the reference for the cell that you want to go to and press `enter` to "jump" to this new location.

Use the shortcut key sequence `ctrl` `G` to initiate the Go To command.

✔ **Press `ctrl` `1`, the equivalent to the End key on a computer.**

Pressing these keys automatically moves you to the last filled cell in a column of data. If you are in an empty column, this feature moves you to the last row (row 2500) of the column.

I often use this feature when I realize that I need to add new values to a current data set.

✔ **Press `ctrl` `7`, the equivalent to the Home key on a computer.**

Pressing this key sequence moves you to row 1 of the current column.

Color Plate 1: Using the new TI-Nspire CX, color images (.jpg, .jpeg, .bmp, and .png files) can be placed as a background to a Graphs page (Chapter 24).

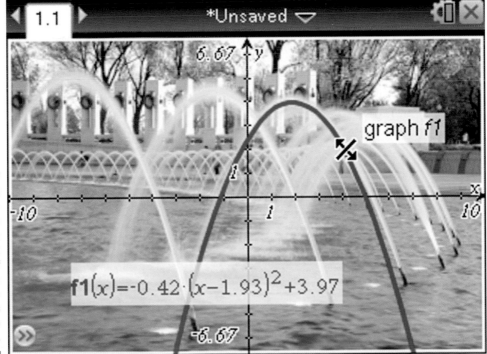

Color Plate 2: Perform a transformation of a graph by taking advantage of the built-in grab & move capabilities of TI-Nspire (Chapter 9).

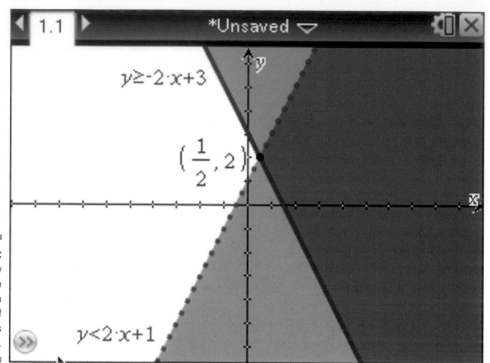

Color Plate 3:
Easily
distinguish the
solution region
of a system of
inequalities
(Chapter 9).

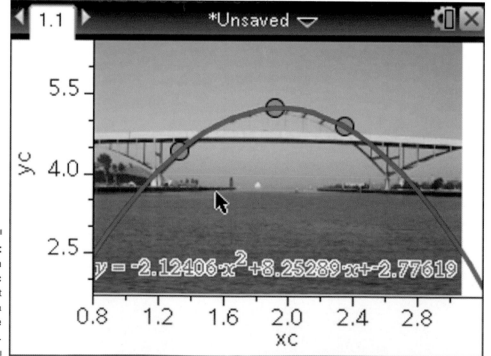

Color Plate 4:
Perform a
quadratic
regression right
on a Data &
Statistics page
(Chapter 19).

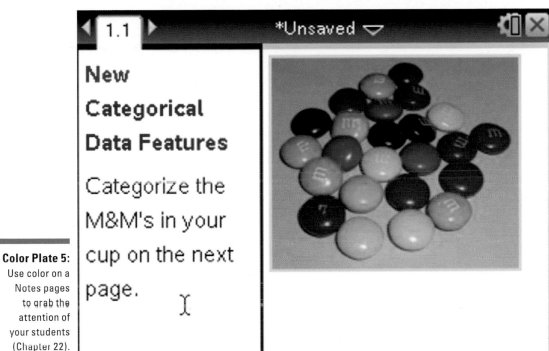

Color Plate 5: Use color on a Notes pages to grab the attention of your students (Chapter 22).

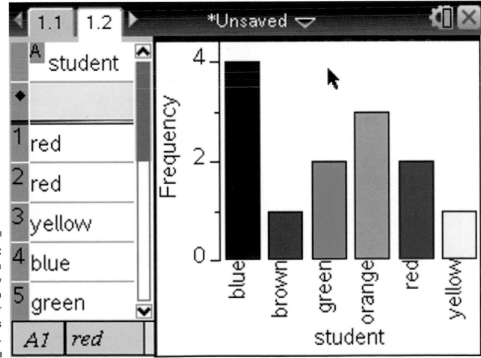

Color Plate 6: Make your data come alive by using color to enhance your bar charts (Chapter 18).

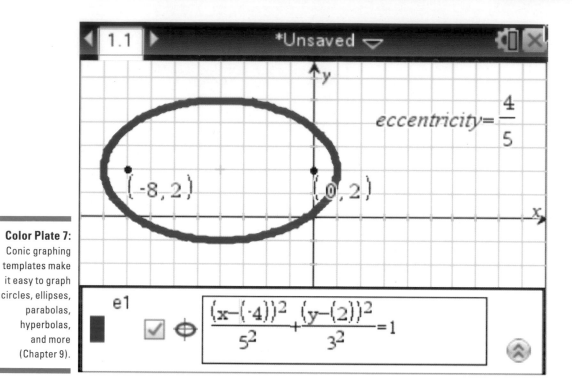

Color Plate 7: Conic graphing templates make it easy to graph circles, ellipses, parabolas, hyperbolas, and more (Chapter 9).

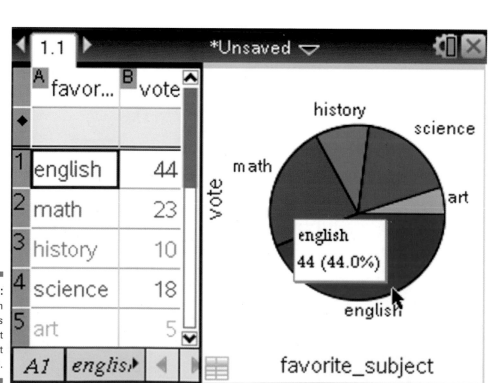

Color Plate 8: Color can even be used in a Lists & Spreadsheet environment (Chapter 13).

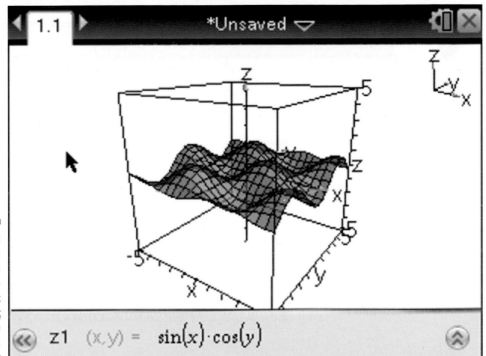

Color Plate 9:
3D graphing allows you to rotate the graph to see it from a different perspective (Chapter 9).

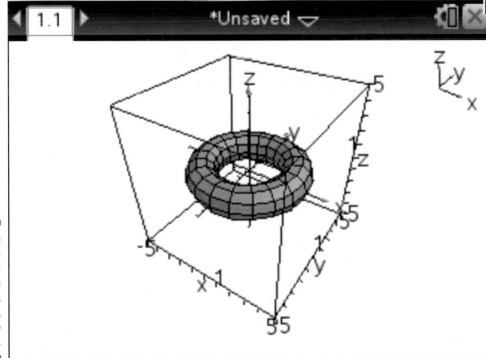

Color Plate 10:
With 3D Parametric graphing, you can graph lines, planes, spheres, and more (Chapter 9).

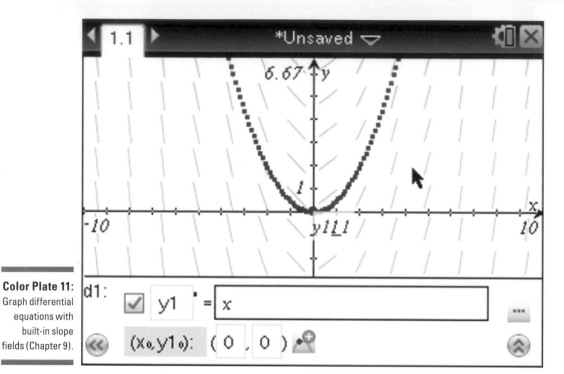

Color Plate 11: Graph differential equations with built-in slope fields (Chapter 9).

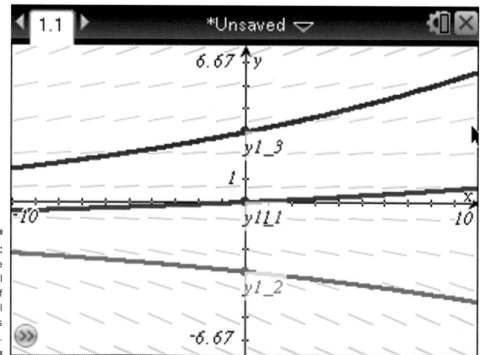

Color Plate 12: Customize the initial conditions of your differential equation graphs (Chapter 9).

Color Plate 13:
Use color in your Polar graphs to make your graphs more dynamic (Chapter 9).

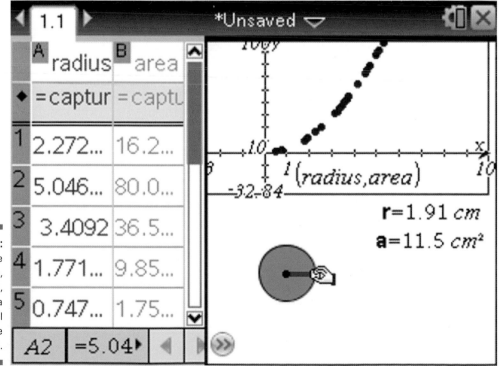

Color Plate 14:
Use multiple representations, capture data, and graph a scatter plot all on one page (Chapter 12).

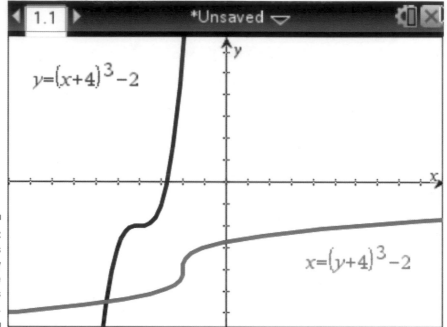

$$y = (x+4)^3 - 2$$

$$x = (y+4)^3 - 2$$

Color Plate 15:
Use text boxes to quickly graph inverse functions (Chapter 9).

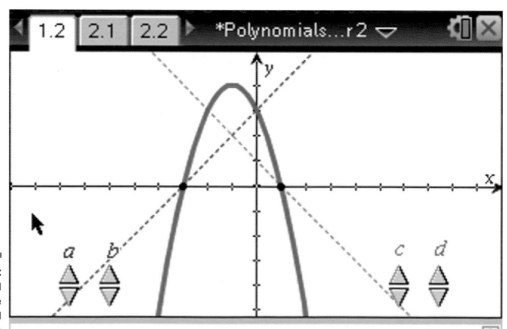

a b c d

Color Plate 16:
Math Nspired activities have been updated to include color features (Chapter 26).

$y_1 = x + 3$ $y_3 = -x^2 - 2x + 3$ $y_2 = -x + 1$

✔ **Press** ⌈ctrl⌉⌈3⌉, **the equivalent to the Page Down key on a computer.**

The TI-Nspire Handheld allows you to view five spreadsheet rows at a time. Therefore, the Page Down key sequence moves you down five rows every time you press these keys.

✔ **Press** ⌈ctrl⌉⌈9⌉, **the equivalent to the Page Up key on a computer.**

Pressing these keys moves you up five rows at a time.

Press ⌈tab⌉ to jump to the column name area located at the top of the column.

Manipulating Rows and Columns

In the following sections, I talk about some of the things you can do to entire rows and columns contained within a spreadsheet.

Resizing columns

In Figure 13-4, notice that the list name *subscription* is truncated. TI-Nspire offers four different options for changing the width of a column:

✔ **Move your cursor to the appropriate column and press** ⌈menu⌉➪**Actions**➪**Resize**➪**Resize Column Width** or right-click to locate the Resize tool. See the first screen in Figure 13-5. After you open this command, press the ◆ keys to set the desired column width and press ⌈enter⌉ or ⌈ ⌉ to lock it in place. See the second screen in Figure 13-5.

✔ **Move your cursor to the appropriate column and press** ⌈menu⌉➪**Actions**➪**Resize**➪**Maximum Column Width.** This action creates a column size of approximately two-thirds of the entire screen!

✔ **Move your cursor to the appropriate column and press** ⌈menu⌉➪**Actions**➪**Resize**➪**Minimum Column Width.** This action reduces the column width to about half of its original size.

✔ **Move your cursor near the top of the columns between column A and column B. When you see the ✛ icon, grab (**⌈ctrl⌉⌈ ⌉**) and use the Touchpad to resize the column as you wish.** This is my favorite way to resize columns, taking advantage of another clickable area on TI-Nspire. See the third screen in Figure 13-5.

It is very rare to need to adjust the height of a row, but you can use another clickable area to accomplish this task. Move your cursor to the beginning of the row (between columns 1 and 2), when you see the ✛ icon, press ⌈ctrl⌉⌈ ⌉ and then press the ▲▼ keys to set the desired row height and press ⌈enter⌉ or ⌈ ⌉ to lock it in place.

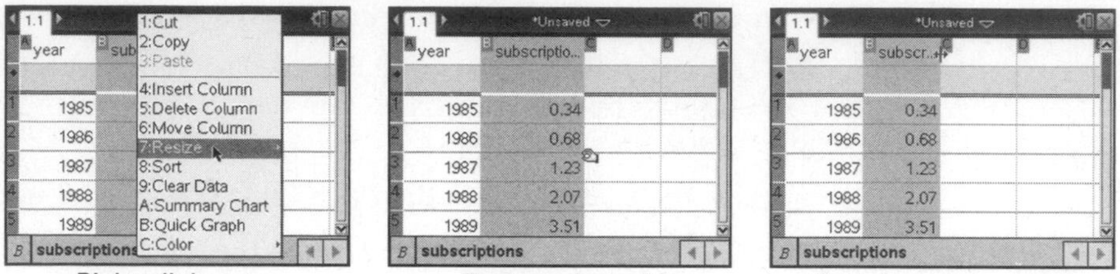

Figure 13-5: Resizing columns.

Right-click menu   Resize column A   Grab to resize column

Moving columns

Consider that you want to move an entire column. To accomplish this task, simply position the cursor in the column that you want to move and press [menu]⇨Actions⇨Move Column to highlight the column (see the first screen in Figure 13-6). Press the ◆ keys to move to the desired location as indicated by the dark vertical line shown in the second screen in Figure 13-6. Press [enter] or [⌖] to set the new position (see the third screen in Figure 13-6).

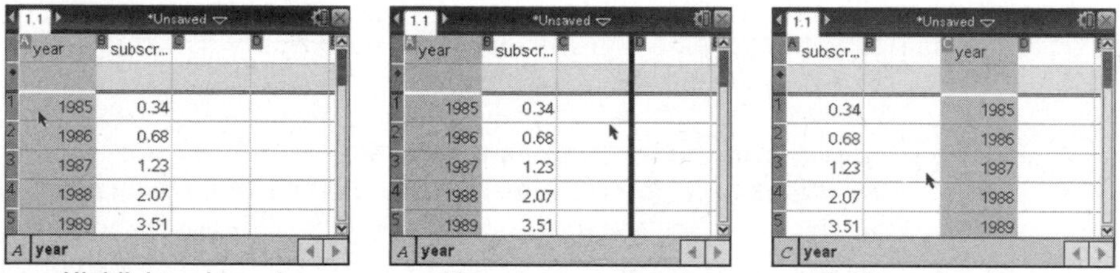

Figure 13-6: Moving columns.

Highlight column A   Move to a new location   Activate the change

Selecting rows and columns

You can perform a variety of editing tasks — such as cutting, copying, or pasting — on entire rows or columns. To do so, you must first highlight the rows or columns with which you want to work. Here's how:

✔ **Select column.** You have three choices to select an entire column:

   • Move the cursor to the column/list name area located at the top of the column and press the ▲ key once more to select the column.

   • Position the cursor anywhere in the desired column and press [menu]⇨Actions⇨Select⇨Select Column.

   • Take advantage of another clickable area. Move your cursor over the capital letter that names the column and click the letter to select the column.

✔ **Select row.** You have three choices to select an entire row:

- Move the cursor to any cell in column A and press the ◄ key once more to select the row.

- Position the cursor anywhere in the desired row and press menu⇨Actions⇨Select⇨Select Row.

- Take advantage of another clickable area. Move your cursor over the number that names the row and click the number to select the row.

✔ **Select multiple rows or columns.** After selecting a row or column, press and hold the ⇧shift key. While holding this key, press the Touchpad keys to select additional rows or columns.

Using Cut, Copy, and Paste

As mentioned in the previous section, highlighted rows and columns can be cut, copied, and pasted. To do this, highlight a row or column and press ctrl X for Cut or ctrl C for Copy, just as you would on a computer. These two actions place the contents of a row or column in temporary memory. Move to an open row or column and press ctrl V to paste these contents.

Deleting and inserting rows and columns

Deleting rows or columns? Nothing could be easier. Just select the row(s) or column(s) to be deleted and press the del key.

Inserting rows and columns? Nothing could be easier, part 2. Press menu⇨Insert⇨Insert Row to insert a row and menu⇨Insert⇨Insert Column to insert a column. Of course, give some consideration to your current cell location. For example, the first screen in Figure 13-7 shows that cell B2 is currently selected. The second screen shows what happens after the Insert Column command is used. The third screen shows what happens after the Insert Row command is used. An empty cell displays as a void, represented by an underscore (_).

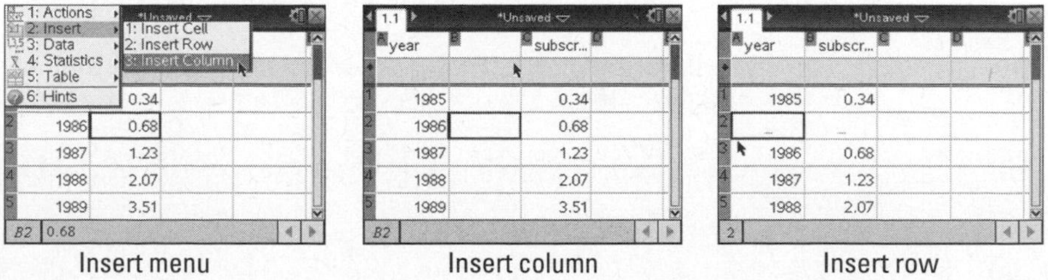

Figure 13-7: Inserting a row or column.

Insert menu   Insert column   Insert row

Using Right-Click

Once again, this amazing shortcut can be a real time-saver. The first screen in Figure 13-8 shows the available options when a single cell is selected. The second screen shows the available options when a single row is selected. The third screen shows the available options when a single column is selected.

Press [ctrl] [menu] to access the right-click context menu.

I encourage you to investigate the options available when a rectangular block of cells is selected as well as when multiple rows and multiple columns are selected.

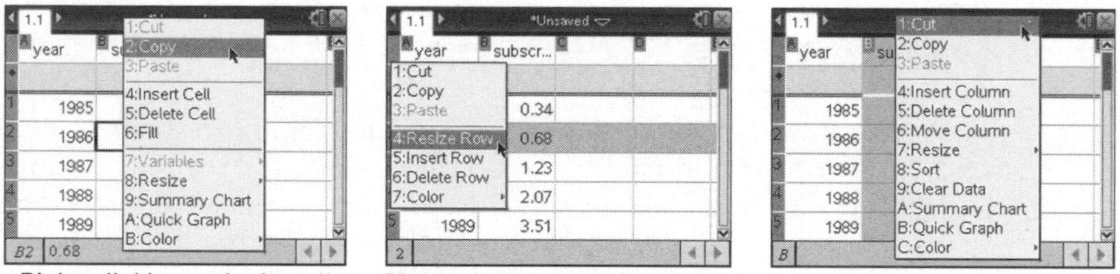

Figure 13-8:
Using the
right-click
option in
Lists &
Spread-
sheet.

Right-clicking a single cell   Right-clicking a selected row   Right-clicking a
selected column

Working with Individual Cells

You have the option of entering numbers, text, and mathematical expressions into a cell. You can also access a variety of formulas and commands from the Catalog menu. The next few sections discuss how to perform these actions.

Entering numbers and typing text

To enter a number in a cell, simply select the cell, type the number, and press [enter] to move to the next cell in the column. Notice that the information you are entering in the cell also appears in the entry line at the bottom of the screen. This is quite helpful if you are entering numbers with several digits.

TI-Nspire deals with text in the very same way. Using the alpha keys, type your text and press [enter].

To enter text as a string, you must use quotes, press [ctrl] [×], and then type your text.

When you enclose text in quotes, your text is *grouped,* allowing words to be referenced elsewhere from within the Lists & Spreadsheet application.

Evaluating mathematical expressions

You can evaluate simple or complex mathematical expressions in the Lists & Spreadsheet application. Just make sure that you press the ☐ key before typing the expression. In Figure 13-9, I show you how to enter an expression (first screen) and show the resulting value of the expression after you press the ⌨ key (second screen). Notice in the second screen of Figure 13-9, I've moved back to cell A1, which reveals the mathematical expression in the entry line at the bottom of the screen.

Highlighting with color

You can use color in the Lists & Spreadsheet application to highlight important parts of the data. Right-click (⌨ ⌨) to access the Color tools. You can highlight a cell by using the Fill Color tool. Alternatively, you can change the color of the text in a cell by using the Text Color tool. See the third screen in Figure 13-9.

Figure 13-9:
Evaluating
an expres-
sion and
using color.

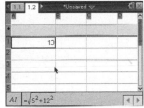

Typing an expression   Evaluating an expression   Highlighting with color

Using the Catalog

To evaluate more sophisticated expressions, press ☐ to access the Catalog. The first screen in Figure 13-10 shows the Catalog. At the bottom of the screen is a syntax example for the highlighted item. The second screen in Figure 13-10 evaluates the random integer command. The last screen in Figure 13-10 incorporates the syntax into the formula row of a spreadsheet, which I discuss in Chapter 14.

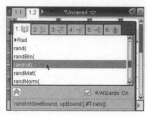 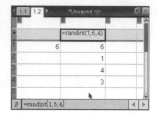

Figure 13-10:
Evaluating
an
expression.

Don't forget to press the ☰ key before entering an expression to be evaluated.

Chapter 14

Working with Data

· ·

In This Chapter

▶ Using cell references to do math

▶ Understanding the difference between relative and absolute cell references

▶ Creating lists of data using the Fill command or from the formula row

▶ Sorting data

▶ Performing statistical analysis on one-variable data

· ·

*I*n this chapter, I get into some of the features of the Lists & Spreadsheet application that are quite similar to those found in other computer-based spreadsheet applications. You see how cell references can be used to quickly perform calculations on a single cell or a rectangular array of cells.

Using Formulas

Imagine that I am interested in converting my height from a value expressed in feet and inches to one expressed in inches only. In the first screen in Figure 14-1, I have typed my height in feet and inches in cells A1 and B1, respectively.

To convert this measurement to inches only, I could move the cursor to cell C1, type =**6*12+3**, and press enter.

Recall from Chapter 13, if you want to perform any calculation in Lists & Spreadsheet, you must first press the ▣ key.

However, instead of typing the number located in a given cell, type the *location* of the number (called the *cell reference*). In the first screen in Figure 14-1, I typed =**a1*12+b1** in cell C1. Pressing enter performs this calculation and displays the result as shown in the second screen in Figure 14-1.

As for my height, I lied. I'm really only 6 feet, 1 inch tall. Because I have used a cell reference, I simply go back to cell B1 and change the 3 to a 1. Notice in the third screen in Figure 14-1 that the calculation in cell C1 updates automatically.

Cell references really come in handy, especially if you are referencing the contents of a cell several times.

Figure 14-1:
Performing
a calcula-
tion using
cell refer-
ences.

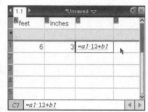  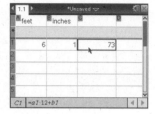

Using relative cell references

Consider now that I want to take the heights of a random group of students and perform the same type of conversion. Here's how:

1. **Type the heights (in feet and inches) in rows 2 through 5.**

2. **Go to cell C1 and press** ⌃ ▣.

   This action selects the cell and allows copying, as indicated by the dashed box. See the first screen in Figure 14-2.

3. **Use the ▾ key to extend the dashed box to cell C5 and press** enter.

   See the second and third screens in Figure 14-2.

Figure 14-2:
Copying a
formula.

Notice that cell C5 is highlighted in the third screen in Figure 14-2 and that the formula is located in the entry line at the bottom of the screen. Recall that the original formula that I typed was =a1*12+b1. The formula has changed to =a5*12+b5. This is because I have used *relative cell references*.

When you copy a relative cell reference, the cell references automatically update. If I copied this formula to cell F7, it would reference the cells directly to the left of this cell and display the formula =d7*12+f7 on the entry line.

Using absolute cell references

Sometimes you want to reference the same cell in a formula, even if you copy the formula to other cells, which is an example of using an *absolute* cell reference. For example, imagine that I want to convert these five heights to centimeters. Follow these steps:

1. **Type the conversion factor for inches to centimeters, 2.54, in cell E1.**

2. **Type the formula** =c1*e$1 **in cell D1 and press** enter.

   The $ symbol, which precedes the row reference, locks the row so that I am always referencing row 1. For example, if I were to copy this formula to cell F7, it would read =e7*g$1. The column is relative and updates accordingly; however, row 1 remains locked. The result of this action is shown in the first two screens in Figure 14-3.

   To access the $ symbol, press ctrl⌨ to open the symbol palette. Scroll down and highlight the $ symbol and press enter to paste this symbol into the formula. You can also press ?!▸ to access the $ symbol.

3. **Copy the formula to cells D2 through D5.**

   Use Steps 2 and 3 from the preceding section to accomplish this task. The last screen in Figure 14-3 shows the result of copying this formula to cells D2 through D5. Notice that the formula for cell D5 is shown on the entry line. Cell C5 is a relative cell reference and was updated when copied. Cell E1 has not been updated because I locked the row reference.

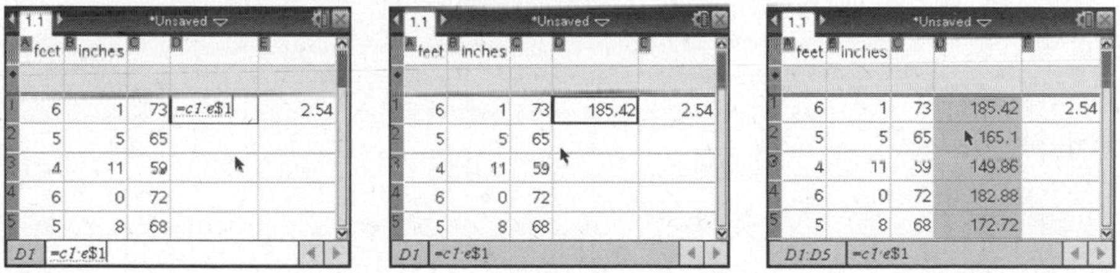

Figure 14-3:
Using an absolute row reference.

The conversion factor that I typed in cell E1 converts inches to centimeters. If I want to perform a different conversion, I can simply change this conversion factor and everything in column D automatically updates. For example, to convert everyone's height to fathoms (for what reason, I'm not sure!), type **0.0139** in cell E1, press enter, and watch the change.

In the previous example, I locked a row. I also have the option of locking both a row and a column. To do so, simply include the $ symbol before both the column reference and the row reference.

In fact, I could have used an absolute row and column reference in this example and accomplished the same result. In the example related to Figure 14-3, the command =c1*e1 references cell E1 no matter where you copy the formula. Similarly, I could have just locked the column reference if the situation called for it.

Referencing a rectangular block of cells

So far, I have shown you how to reference individual cells. I can also reference rectangular blocks of cells. To do this, type the cell reference for the upper-left corner of the rectangular block of cells, press ⌨ to select :(colon), and type the cell reference for the lower-right corner of the rectangular block of cells.

For example, I have two classes of Algebra II, one that meets during Period 2 and one that meets during Period 3. I'd like to calculate the mean (average) for each class as well as the combined mean for both classes. The grades for Period 2 are located in column A and the grades for Period 3 are located in column B.

You have three ways to accomplish this task:

✔ **Place the cursor in cell D2 and type** =mean(a1:a5).

The cell shows the mean for Period 2.

✔ **Place the cursor in cell D3 and type** =mean(b1:b5)*1.0.

The cell shows the mean for Period 3.

I forced the result in cell D3 to be displayed as a decimal (rather than a fraction) by multiplying by 1.0. Anytime you include a decimal in a calculation, the result is displayed in decimal form.

✔ **Place the cursor in cell D4 and type** mean(a1:b5)*1.0.

The cell shows the mean for both classes. See the first screen in Figure 14-4.

I can also use the Select Range tool to name a block of cells. Using this tool is similar to selecting a block of cells using a spreadsheet in computer software:

1. **Place the cursor in cell D4 and type** =mean(.

2. **Press** ⟨menu⟩⇨**Actions**⇨**Select**⇨**Select Range to invoke the Select Range tool.**

   A dotted selection rectangle appears around cell D4.

3. **Move to cell A1 and hold** ⟨⇧shift⟩ **while pressing the Touchpad keys.**

   As you move, the dotted selection rectangle encloses the cells that you would like to select for your calculation. See the second and third screens in Figure 14-4.

4. Press 〔enter〕 to evaluate and display the result.

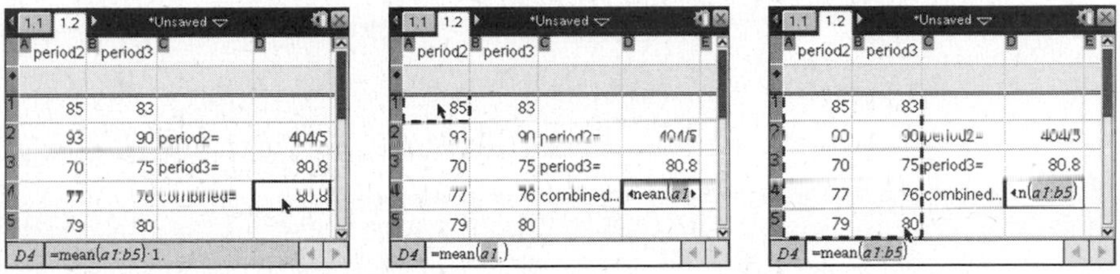

Figure 14-4:
Referencing
a rectangu-
lar block of
cells.

You can also access the Mean command from the Catalog (〔🔖〕).

Working with Data

TI-Nspire allows you to generate lists of sequential data very quickly. And the analysis of these data sets can be performed quickly and efficiently as well. Again, if you are familiar with spreadsheets, you can draw on this past experience. If you are not, I have you covered.

Using the Fill command

A *recursive formula* is a formula that is used to determine the next term in a sequence by using one or more of the preceding terms. The Fill Down command can be used to generate a recursive sequence in the case where you always use the previous term to obtain the next term.

Consider that you want to explore the future value of a $1,000 deposit under two different scenarios. Under the first scenario, you earn 5 percent annual interest. Under the second scenario, you simply add $100 each year to the account. Here's how it's done.

1. **Type the initial deposit, 1000, in cells A1 and B1.**

2. **Type =a1*1.05 in cell A2 and =b1+100 in cell B2.**

3. **Move the cursor to cell A2 and press 〔menu〕⇨Data⇨Fill to grab the cell as indicated by the dashed box.**

   Here are two other options for accessing the Fill command:

   • Move the cursor to cell A2 and press 〔ctrl〕〔📋〕 (only two keystrokes!).

   • Right-click in cell A2, 〔ctrl〕〔menu〕⇨Fill.

4. **Press the ▾ key repeatedly to expand the dashed box to the desired row and press** [enter].

5. **Repeat Steps 3 and 4 for cell B2.**

In Figure 14-5, I have used the Fill command to copy my formulas to row 10.

Notice that the value of the account in column B is higher than the account value in column A. Of course, I'd like to find out when the first investment scenario exceeds the second.

To use Fill on both columns simultaneously, move to cell A10, press and hold the [⇧shift] key, and press the ▸ key once to highlight cells A10 and B10. Now access the Fill command as before and copy both formulas to several more rows (I've filled down to row 30). Scroll down to see that the 5% investment strategy surpasses the other investment strategy in row 28!

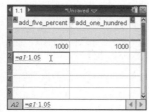
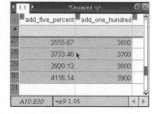

Figure 14-5: Using the Fill command.

The Fill Command can also be used horizontally to fill in a row of values.

Using the formula row

The formula row is the row that is located just above row 1. It is denoted by the ♦ located at the left side of the row. In the following sections, I show how the formula row can be used to manage large data sets without the need to work within the cells themselves. Each column has one formula cell that is located in the formula row.

Generating sequential data

One of my favorite sequences is the famous Fibonacci sequence. The first two values of this sequence are both 1. The remaining values are found by adding the two previous values. To create this sequence, move the cursor to the formula cell for column A and follow these steps:

1. **Press** [menu]⇨**Data**⇨**Generate Sequence or right-click,** [ctrl][menu]⇨**Generate Sequence.**

See the first screen in Figure 14-6.

2. **Configure the dialog box as shown below.**

   Formula: u(n)=u(n-2)+u(n-1). **Initial Terms:**1,1. **n0:**1. **nMax:**255. **nStep:**1. **Ceiling Value:**100. See the second screen in Figure 14-6.

3. **Press [enter] to generate the sequence.**

   See the third screen in Figure 14-6.

Notice that I have two options for controlling the number of terms in my sequence — by maximum number of terms and by ceiling value. If you want to specify the maximum number of terms, say 20, just type the number in the Max No. Terms field and leave Ceiling Value blank. If you type a maximum value in the Ceiling Value field, TI-Nspire defaults to the Ceiling Value and ignores the Max No. Terms field.

Figure 14-6:
Using the
Generate
Sequence
command.

Right-click menu Sequence dialog box Fibonacci sequence

Perhaps I want to edit the sequence command to include more terms. Just move to the formula cell and press 🔢 or [enter] to enter Edit mode. Then move the cursor to the end of the command by pressing the ▸ key repeatedly (or press [ctrl][1], the equivalent of the End key on a computer). Delete the last number and type a new number.

Working with random numbers

I find that the random numbers and the Lists & Spreadsheet application are a perfect match.

Consider that I want to simulate rolling two dice 50 times using the Random Integer command. Here's how:

1. **Move to the formula cell for column A, type** =randint(1,6,50), **and press [enter].**

   You can type the randint command using the green alpha keys or access the randint command from the Catalog by pressing the 📖 key.

2. **Repeat Step 1 for column B.**

Notice that columns A and B are filled with 50 random integers from 1 to 6.

The syntax for the Random Integer command is

```
randInt(lower bound, upper bound[, number of trials])
```

Creating column data based on another column

Now I'm interested in finding the sum of the dice for each trial. I can use the header row to accomplish this task.

Just move to the formula cell for column C, type **=a[]+b[]**, and press ⏎. The first screen in Figure 14-7 shows the results of this action.

When referencing a column in the formula row, you must include an empty set of brackets, accessed by pressing ctrl (after the column letter.

Perhaps I chose to name the first three columns *first_die, second_die,* and *total,* respectively. Rather than use a letter reference for a column, I can reference these list names. After naming the first two columns (see Chapter 13), follow these steps:

1. **Move the cursor to the formula cell for column C and press the ⚌ key.**

   Notice that `total:=` automatically appears in the header row.

2. **Press** var **, highlight *first_die*, and press** enter **.**

3. **Press** + **.**

4. **Press** var **, highlight *second_die*, and press** enter **.**

   See the second screen in Figure 14-7.

5. **Press** enter **to complete the command.**

   See the third screen in Figure 14-7.

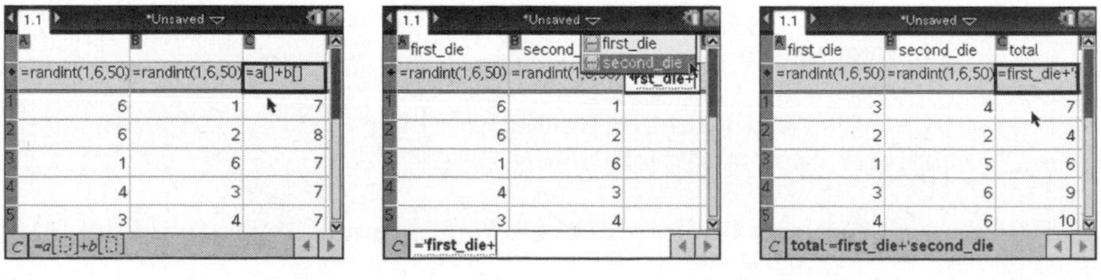

Figure 14-7:
Creating
column data
based on
other
columns.

If you are following along with me, your screen may differ from the screen shown in Figure 14-7. That's okay. Remember, these are random integers and it's to be expected that they will differ each time.

Recalculating cell references and formulas

You can update all cell references and formulas by pressing menu⇨Actions⇨ Recalculate. Alternatively, press ctrl R, the shortcut method for accessing the Recalculate command.

The Recalculate command is particularly handy when working with random numbers. Each time that you press ctrl R, you get an entirely new set of random numbers. In the case of the two-dice experiment, it's like rolling two dice another 50 times each time you use this command.

Sorting data

Perhaps you want to sort the data that's contained in column C from the two-dice experiment to see how the sums are distributed. Follow these steps and refer to the screen images in Figure 14-8.

1. **Select columns A, B, and C.**

   Move the cursor to the column/list name area located at the top of the column and press the ▲ key once to select the column. While holding the ⇧shift key, press the Touchpad keys to select the additional columns.

2. **Press menu⇨Actions⇨Sort (alternatively, right-click by pressing ctrl menu⇨Sort).**

   At the warning prompt, press enter.

3. **Press ⬆ to see the choices for Sort By, and select c.**

4. **Press enter to sort by Ascending or highlight this field and change it to Descending and press enter.**

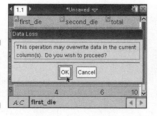

Figure 14-8: Sorting data.

Notice in the third screen in Figure 14-8, the formulas in the formula row have been deleted. The formulas will be deleted anytime you do a sort on a column containing a formula. Your random numbers are now static and `recalculate` no longer gives you a new set of random numbers. Pressing ctrl esc undoes the sort and restores all the formulas in the formula row.

If you select only column C and do a sort, the relationship among columns A, B, and C is lost. Each value in column C is likely not to equal the sum of the two numbers to its left.

Basic Statistical Analysis

In the following sections, I show you some basic statistical functions that can be performed on a single data set. In Chapter 15, I show you how to perform statistical analyses on two-variable data sets.

Using one-variable statistics

I've entered the first-semester averages for my two Algebra 2 classes in a column titled *algebra2,* and I'm interested in performing a one-variable statistical analysis on this data. Here are the steps:

1. **Press** menu⇨**Statistics**⇨**Stat Calculations**⇨**One-Variable Statistics.**

2. **Press** enter **to indicate that you want to analyze one list.**

   If you have additional lists, change the Num of Lists field to match the number of lists that you are interested in analyzing.

3. **Configure the dialog box as shown in Figure 14-9.**

Figure 14-9:
Performing
one-variable
statistical
analysis.

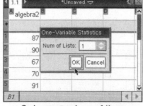
Select number of lists

Configure dialog box

One-variable
statistical results

Scroll down column C to review the statistical results that are generated by this action. The following list contains a description of what each result means:

- ✔ x̄: Sample mean
- ✔ Σx: Sum of the data
- ✔ Σx²: Sum of the squared data

- **sx:** Sample standard deviation
- **σx:** Population standard deviation
- **n:** Sample size
- **MinX:** Minimum value
- **Q$_1$X:** First quartile
- **MedianX:** Median
- **Q$_3$X:** Third quartile
- **MaxX:** Maximum value
- **SSX:** Sum of squared deviations

I also could have created a separate list for each Algebra II class, one in column A and one in column B. To do this, press menu⇨Statistics⇨Stat Calculations⇨Two-Variable Statistics to obtain the single-variable statistics for both classes simultaneously.

Statistical analysis using the Catalog

You also have the option of finding individual statistics on a data set by accessing commands from the Catalog (⌨).

In the first and second screens in Figure 14-10, I've accessed the Mean function from the Catalog. In the third screen, I accessed the Sum command. Notice that I must press = before inserting a command. I also pressed var and chose algebra2 to paste my list name within each of these commands.

Figure 14-10: Using the Catalog to find statistical results.

I can type a function using the alpha keys rather than access it via the Catalog.

Quickly copying data from the Internet into your TI-Nspire

Data from the Internet can be transferred to your TI-Nspire in an easy, two-step process. To demonstrate this process, I used my computer to access the 2009 regular season quarterback stats on the Nfl.com Web site. At the top of the ratings is the fabulous Drew Brees (whose team, the New Orleans Saints, won the Super Bowl in 2009). The last quarterback in the ratings is JaMarcus Russell, who threw three touchdowns and 11 interceptions (ouch!). The following figure shows a sampling of the 640 pieces of data that I gathered from the site at http://www.nfl.com/stats/category stats?tabSeq=1&statisticPositi onCategory=QUARTERBACK&season= 2009&seasonType=REG.

| All NFL ▾ | Passing ▾ | 2009 ▾ | Regular Season ▾ | All ▾ | Go |

Passing (Qualified¹ | All)

Rk	Player	Team	Pos	Comp	Att	Pct	Att/G	Yds	Avg	Yds/G	TD	Int	1st	1st%	Lng	20+	40+	Sck ▾	Rate
1	Drew Brees	NO	QB	363	514	70.6	34.3	4,388	8.5	292.5	34	11	210	40.9	75T	58	11	20	109.6
2	Brett Favre	MIN	QB	363	531	68.4	33.2	4,202	7.9	262.6	33	7	211	39.7	63	52	13	34	107.2
3	Philip Rivers	SD	QB	317	486	65.2	30.4	4,254	8.8	265.9	28	9	208	42.8	81T	64	12	25	104.4
4	Aaron Rodgers	GB	QB	350	541	64.7	33.8	4,434	8.2	277.1	30	7	197	36.4	83T	55	17	50	103.2
5	Ben Roethlisberger	PIT	QB	337	506	66.6	33.7	4,328	8.6	288.5	26	12	202	39.9	60T	61	14	50	100.5
6	Peyton Manning	IND	QB	393	571	68.8	35.7	4,500	7.9	281.2	33	16	237	41.5	80T	59	8	10	99.9
7	Matt Schaub	HOU	QB	396	583	67.9	36.4	4,770	8.2	298.1	29	15	230	39.5	72T	62	15	25	98.6
8	Tony Romo	DAL	QB	347	550	63.1	34.4	4,483	8.2	280.2	26	9	203	36.9	80T	61	17	34	97.6
9	Tom Brady	NE	QB	371	565	65.7	35.3	4,398	7.8	274.9	28	13	214	37.9	81T	43	12	16	96.2
10	Kurt Warner	ARI	QB	339	513	66.1	34.2	3,753	7.3	250.2	26	14	191	37.2	45	42	3	24	93.2

The first step is to copy the data from the Internet. Simply highlight the data that you want and press Ctrl+c to copy the data.

Next, add a Lists & Spreadsheet page to a document using TI-Nspire Teacher Software. Place your cursor in cell A1 and press Ctrl+v to paste the data. See the first image in this figure. The column headers will not transfer to the column list name area on the TI-Nspire. Instead, they will be located in the first row. In the second screen, I typed the column headers in the column list name area of each column. The last screen in the figure shows a box plot of the quarterback ratings. I inserted a Data & Statistics page to create the box plot.

Finally, you will need to transfer the .tns file from the computer software to your handheld. You can learn the steps to make the transfer in Chapter 25.

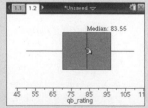

| Paste the data in L&S | Name the column | Create a box plot |

Chapter 15

Constructing Scatter Plots and Performing Regressions

. . . .

In This Chapter

▶ Entering data in the Lists & Spreadsheet application

▶ Constructing a scatter plot of data

▶ Inspecting a scatter plot and performing a suitable regression

▶ Graphing and updating a regression equation

▶ Working with more than one regression

. . . .

Finding a mathematical model for a data set is a common application found in a typical algebra or higher-level math course. Mathematical models are very important because they enable us to create an analytical representation of numerical data. In turn, this mathematical model (or equation) can be used to make predictions for values of the input variable for which data is not available.

In this chapter, I focus on how to input data into the Lists & Spreadsheet application, inspect the corresponding scatter plot in the Graphs application, and find an appropriate model to fit the data using TI-Nspire's regression capabilities. I then show you how to graph your regression equation along with the scatter plot. Finally, I talk about the dynamic capability of TI-Nspire to update results from a regression without performing the regression a second time.

Constructing a Scatter Plot in the Graphs Application

The phrase "a picture is worth a thousand words" rings quite true when you are trying to find a mathematical equation to model a data set. In fact, before you perform a regression, it's almost imperative that you view a scatter plot of the data to assist you in deciding on an appropriate model. Yes, you have ways of looking at the numerical data to determine a good model (such as finding first- and second-order differences); however, an inspection of a scatter plot can often reveal this information much more efficiently.

Consider that you are interested in looking at the following relationship:

- ✔ United States immigrant population (in millions) versus years after 1900
- ✔ Total United States population (in millions) versus years after 1900

Table 15-1 gives the numerical data for both these relationships.

Table 15-1	United States Immigrant Population Since 1900	
Years Since 1900	*Immigrant Population (in millions)*	*U.S. Population (in millions)*
0	10.3	76.2
10	13.5	92.2
20	13.9	106
30	14.2	123.2
40	11.6	132.2
50	10.3	151.3
60	9.7	179.3
70	9.6	203.2
80	14.1	226.5
90	19.8	248.7
100	31.1	281.4

Entering the data

Here are the steps to enter the data in a Lists & Spreadsheet page:

1. **To start with a new document, press** ctrl N **(a shortcut key sequence) or press** 🏠on ⇨**New Document.**

   If you currently have an open document, you are prompted to save the document. Press enter for Yes or press tab enter for No. I usually tell my students to "just say No" when prompted to save changes.

2. **On the next screen, select Add Lists & Spreadsheet and press** enter.

   Figure 15-1 shows this sequence of screens.

  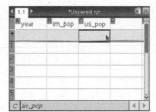

Figure 15-1: Opening a new Lists & Spreadsheet page.

Save file prompt   Select application   Blank L & S screen

Before entering your data, it's a good idea to name your lists.

It's always a good idea to choose names that convey the meaning of the data the lists represent because the list names show up as the *x*- and *y*-variables of the scatter plot

3. **Use the Touchpad to move your cursor to the column list name area.**

4. **Type your list name using the alpha keys and press ⌶tab⌷ to move to the column list name area in column B.**

   Again, type a name that describes the immigrant data, and then move to column C and choose a meaningful name for the U.S. population data. Because I am not allowed to use a space when naming a column, I like to use an underscore (press ?!▸ or ctrl ⌂). In Figure 15-2, you can see that I named the lists: *year, im_pop,* and *us_pop.*

Figure 15-2: Naming each column that contains data.

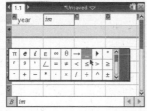 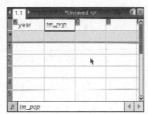

5. **Enter the data in column A.**

   Notice that the data for *year* forms the arithmetic sequence 0, 10, 20, . . . , 100. Here are three time-saving options for entering this data that eliminate the need to actually type each number:

   - **Use the Fill command with a formula.** Type the number 0 in cell A1. Next, type the command =**a1+10** in cell A2 and press ⌶enter⌷. Position the cursor back in cell A2 and press ⌶menu⌷⇨Data⇨Fill. Notice that the box around cell A2 becomes dashed, indicating that it has been selected. Press the ▾ key repeatedly until the dashed box reaches row 11 and press ⌶enter⌷. The third screen in Figure 15-3 shows that the sequence is complete.

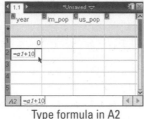

Figure 15-3:
Using the Fill
command to
generate a
sequence.

Type formula in A2   Grab cell A2   Fill down

- **Use the Fill command without a formula.** Type **0** in cell A1 and
  10 in cell A2. Next, position the cursor in cell A1, press the 〈shift〉
  key, and press ▼ once to highlight cells A1 and A2, as shown in the
  first screen in Figure 15-4. Press 〈ctrl〉〈menu〉⇨Fill. Then press the ▼ key
  repeatedly until the dashed box reaches row 11, and press 〈enter〉. The
  second screen in Figure 15-4 shows that the sequence is complete.

This method of highlighting two cells to generate a sequence
works only for a sequence that is arithmetic, meaning that each
term is generated by adding or subtracting the same value from
the previous term. If you want to create a different sequence, such
as the geometric sequence 1, 2, 4, 8, . . . , 128, use Fill with a formula or the method shown in the next bullet.

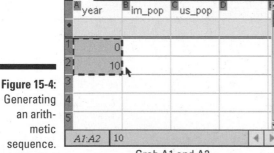

Figure 15-4:
Generating
an arith-
metic
sequence.

Grab A1 and A2   Fill down

- **Use the Sequence command.** Position the cursor in the column
  A formula cell (the gray box located above row 1) and press
  〈▥〉. Press 〈S〉 to quickly jump to the commands that begin with
  the letter S, and then locate the sequence command using the
  Touchpad keys. (See the first screen in Figure 15-5.) Use the syntax
  example at the bottom of the Catalog page to help you enter the
  sequence correctly. I chose to configure a recursive sequence by
  typing **=seq(10(n-1),n,1,11)**. Press 〈enter〉 to execute the command
  and populate cells A1 through A11 with the values 0, 10, 20, . . . ,
  100. (See the third screen in Figure 15-5.)

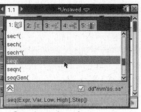 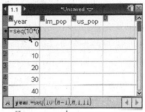

Figure 15-5:
Generating
a sequence
using the
Sequence
command.

Catalog showing syntax   Using the *seq* command   Generate the sequence

6. Enter the data for columns B and C.

The data for column B, U.S. Immigrant Population (in millions), and that
of column C, U.S. Population (in millions), must be entered manually.
Refer to Table 15-1 and enter the data one value at a time.

Insert a Graphs page

You are now ready to view the scatter plot of *U.S. immigrant* versus *year*
population. Here's what you do:

1. **Press** ctrl I **and select Add Graphs. Alternatively, press** 🏠on **and select
   Graphs.**

   Notice that the cursor is in the entry line located at the bottom of the
   screen, indicating that the entry line is currently active. Also note that
   the entry line is set by default for graphing functions.

2. **Press** menu⇨**Graph Type**⇨**Scatter Plot to switch the entry line to scatter
   plot. Alternatively, press** ctrl menu ⇨**Graph Type**⇨**Scatter Plot while the
   cursor is located in the entry line. Remember, this second option is
   the equivalent of a right-click on a computer.**

   In the first screen in Figure 15-6, I have pressed var to view a list of
   choices for the *x*-variable.

3. **Using the** ▲▼ **keys, highlight** *year* **and press** 🔄 **or** enter.

4. **Press** tab **to move to the** *y*-variable, **press** var **to view a list of choices
   for the** *y*-variable, **highlight** *im_pop,* **and press** 🔄 **or** enter. **Press** enter **to
   graph the scatter plot.**

If you configured your scatter plot with incorrect list names, you can press tab
to move from the work area back to the entry line. Notice that the next available unused scatter plot (probably *s2*) is ready to be configured. To edit the
previous scatter plot, press ▲ and change the settings as needed.

Notice in the last screen of Figure 15-6 that the current window settings do
not provide a good view of the scatter plot. In fact, you see no visible points
in the viewing window.

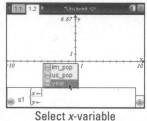
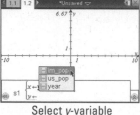

Figure 15-6:
Configuring
Graphs to
display a
scatter plot.

| Select *x*-variable | Select *y*-variable | Graph scatter plot |

To improve the view of the scatter plot, press menu⇨Window⇨Zoom–Data. This sets your window settings to match those shown in Figure 15-7.

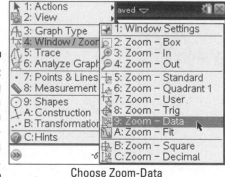
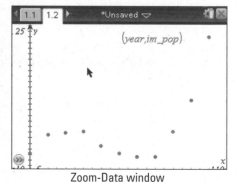

Figure 15-7:
Using
Zoom–Data
to set an
appropri-
ate viewing
window.

| Choose Zoom-Data | Zoom-Data window |

REMEMBER

The shortcut key sequence ctrl G toggles between Hide Entry Line and Show Entry Line.

If you inspect the scatter plot, it appears that a cubic polynomial is a good fit for the data. In this next section, I tell you how to perform this type of regression.

Performing a Regression

To perform a regression, follow these steps:

1. **Press ctrl ◄ to move back to the Lists & Spreadsheet page containing the data for *year* and *U.S. immigrant population*.**

2. **Press menu⇨Statistics⇨Stat Calculations⇨Cubic Regression.**

   A dialog box opens, as shown in Figure 15-8. As with any dialog box, you can press tab to move from one field to the next or ⇧shift tab to move backward through the fields. Press enter at any time for the configuration to take effect.

Notice in the first screen in Figure 15-8 that I have pressed ⬚ and selected *year* to specify the location of the X List. I could also specify the column by its letter reference, which is what I did for the Y List (by pressing ⬚ ctrl ⬚). In the second screen in Figure 15-8, I have pressed ⬚ ctrl ⬚ to indicate where I want the results to be stored. I selected column D because it is the first available empty column.

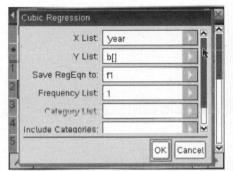 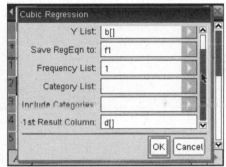

Configuring the Cubic Regression dialog box

Figure 15-8:
Performing
a cubic
regression.

Cubic regression results

The third and fourth screens in Figure 15-8 (with columns D and E widened) show that the results of the regression are pasted directly into the spreadsheet. Notice that the *coefficient of determination*, R^2, is very close to 1. This might suggest that the choice of doing a cubic regression was a good one.

A cubic regression can be done with a minimum of four points. Under this scenario, TI-Nspire finds a third-degree polynomial fit by solving a system of four equations determined by using the *x*- and *y*-values of the four points. With five or more points, TI-Nspire performs a quartic regression.

Understanding the results of a regression

A variety of variables (26 in all) are stored by TI-Nspire after a regression. To view this list, add a Calculator page and press $\boxed{\text{var}}$. See the first screen in Figure 15-9. Using the ▲▼ keys, you can scroll through this list and paste a variable into the entry line on the Calculator page.

For example, I've pasted the variable stat.resid to the entry line (see the second screen in Figure 15-9) and pressed $\boxed{\text{enter}}$ to view a list of *residuals*. This list represents the difference between the *y*-value of each data point and the corresponding *y*-value associated with the regression equation.

It might be interesting to inspect a scatter plot of the residuals as a function of the year. This plot visually shows the difference between the *y*-value of each data point and the corresponding *y*-value associated with the regression equation. To construct this scatter plot, use the same methods just described. Let *year* be the variable used for X List and stat.resid be the variable used for Y List. The graph of this scatter plot is shown in the third screen of Figure 15-9. A perfect-fit regression places the residuals directly on the *x*-axis because no variation exists between the regression model and the data points.

Figure 15-9:
Viewing residuals on a Calculator page and the related scatter plot.

Regression variables

Residuals list

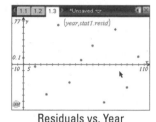
Residuals vs. Year

Referring to the first screen of Figure 15-9, you see several other statistical variables generated from this cubic regression. For example, the variable stat.results, when executed on a Calculator page, displays a list of all the regression results pasted into columns D and E when the cubic regression was first performed on the Lists & Spreadsheet page.

Graphing your regression equation

To view the graph of your regression equation, follow these steps:

1. **Move back to the Graphs page. If you hid the entry line earlier, press** ⌃ G **to bring it back into view. Also, change the entry line to Function by pressing** ⌃ menu ⇨**Graph Type**⇨**Function or** menu ⇨**Graph Type**⇨**Function.**

   Recall that you configured the regression dialog box to save the regression equation to f1.

2. **Press ▲ until your regression equation appears, as shown in the first screen in Figure 15-10.**

3. **Next, press** enter **to activate the graph of the regression equation, as shown in the second screen in Figure 15-10.**

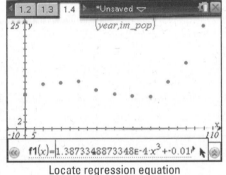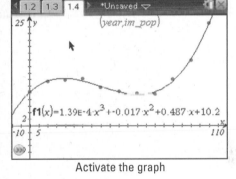

Figure 15-10:
Graphing a
regression
equation.

Locate regression equation   Activate the graph

Updating your regression equation

The data contained in Table 15-1 gives population data for the years 1900 through 2000. Imagine that you want to add more-recent data. For example, the U.S. immigrant population in the year 2006 was approximately 37.5 million. TI-Nspire allows you to add or revise data and then update the regression automatically. Here is what you do:

1. **Move your cursor to column A and press** ⌃ 1 **to jump down to the bottom of the list.**

2. **Type** 106 **in cell A12 and** 37.5 **in cell B12.**

3. **Move back to the Graphs page and compare the results shown in Figure 15-11 with those from the original regression in Figure 15-10.**

   Although the change is subtle, you can clearly see that the scatter plot and regression equation have changed. Also, take a look at the regression results located in columns D and E of the Lists & Spreadsheet page. The regression results have been updated there as well.

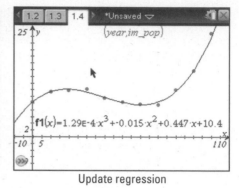

Figure 15-11:
Updating a
regression.

Update data   Update regression

If you use the Sequence command in the column A formula cell to generate the X List data and then try to add data at the bottom of the list, you see the following warning:

> This operation will overwrite data in the current column. Do you wish to proceed?

Pressing [enter] (indicating yes) deletes the existing column data and replaces it with zeros. Instead, use the Fill method for generating the first 11 terms of the sequence. Then, manually add 2006 to the list.

Performing a Second Regression

Perhaps you want to construct a scatter plot and associated regression for *Total U.S. population* versus *year*. This shouldn't be too hard. Just follow these steps, which also serve as a nice summary for this chapter:

1. **Enter the data.**

   If you haven't already done so, refer to Table 15-1 and enter the data for U.S. Population in column C.

2. **Configure a second scatter plot.**

   On the Graphs page, configure a second scatter plot (*s2* in this case) to plot *year* for *x* and *us_pop* for *y*. Make sure that you delete the value in cell A12; otherwise, you will get an error message. The dimensions of two lists must be the same, or you cannot perform a regression.

   Because you are viewing a new scatter plot, the window settings must be adjusted.

3. **Change the window settings. Press** [menu]⇨**Window**⇨**Zoom – Data. See the first screen in Figure 15-12.**

4. Decide on a regression model.

Although the scatter plot might suggest a linear model, what you know about population growth also suggests that an exponential model might be a better fit.

5. Perform an exponential regression:

a. Go back to the Lists & Spreadsheet page and press menu⇨Statistics⇨Stat Calculations⇨Exponential Regression.

b. Configure the dialog box as shown in the second screen in Figure 15-12, and press enter to view the results of the regression, as shown in the third screen in Figure 15-12.

6. Graph the regression equation. Go back to the Graphs page and plot your regression equation (stored in *f2*) as shown in the last screen in Figure 15-12.

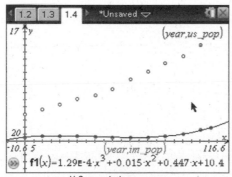

U.S. population vs. year

Configure the dialog box

Figure 15-12: Performing a second regression.

Perform regression

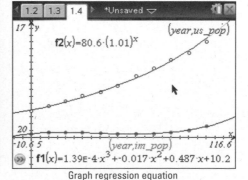

Graph regression equation

Notice that in Figure 15-12, the value of the correlation coefficient, *r*, is 0.9975. . . . Had you performed a linear regression, the value of *r* would have been slightly lower (0.9906 . . .). This suggests that the exponential regression provides a better fit than a linear regression.

You may have noticed that the second scatter plot consists of hollow points to offset them from the first scatter plot. You can change the attributes of a scatter plot by moving the cursor on the scatter plot and pressing [ctrl] [menu]⇨Attributes. In total, you have nine different scatter plot styles to choose from, as shown in Figure 15-13.

Figure 15-13: Scatter plot style choices.

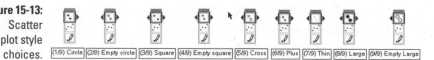

Chapter 16

Manual and Automatic Data Capture

- -

In This Chapter

▶ Setting up an activity that uses the Data Capture feature

▶ Storing variables to be captured

▶ Configuring the Lists & Spreadsheet application for data capture

▶ Setting up the Geometry application to view captured data

▶ Performing a data capture experiment and analyzing the results

- -

*I*n case you haven't already noticed, TI-Nspire is unique in its capability to show multiple representations of mathematical concepts. Furthermore, TI-Nspire's grab-and-move functionality allows you to view these representations dynamically. If I had to choose one aspect of TI-Nspire that embodies these features, it would be the Data Capture feature.

In this chapter, I show you a classic math problem that can be analyzed using manual and automatic data capture. You will also see how data capture is a perfect fit for use with both the Geometry application and the Lists & Spreadsheet application.

Storing Variables in Geometry

Many situations arise in mathematics that require you to investigate a geometric object numerically and algebraically. Take, for example, the situation in which you want to determine the maximum area enclosed by a rectangle with a fixed perimeter.

To analyze this situation, open a Geometry page and press menu⇨View⇨Show Analytic Window. Press tab until the word *axes* appears to change the size of the analytic window. See the first screen in Figure 16-1. Set your analytic window to have the same window settings as those shown in the second screen. Add a separate page with a Lists & Spreadsheet application. See the third screen.

Figure 16-1:
Adding
Geometry
and Lists &
Spread-
sheet
pages.

Tab to the axes   Adjust the analytic window   Insert L & S page

You are now ready to construct a rectangle and obtain some measurements. Here are the steps to follow:

1. Press [menu]⇨Shapes⇨Rectangle to access the Rectangle tool.

   a. Move the cursor to place the upper-left corner of the rectangle and press [🖰].

   b. Press [⇧shift], then move the cursor to place the upper-right corner of the rectangle and press [🖰]. (Pressing the Shift key allows for movement in 15-degree increments.)

   c. Move the cursor to place the lower-right corner of the rectangle and press [🖰].

2. Press [menu]⇨Measurement⇨Length to open the Length tool.

   a. To measure the perimeter of the rectangle, move the cursor to the rectangle until you see `rectangle` [tab]. Press [🖰] to measure the perimeter of the rectangle. Move the measurement using the Touchpad keys and press [🖰] to drop the measurement in place.

   b. Move the cursor to the side that represents the length of the rectangle (I chose the bottom) until you see `rectangle` [tab] again. Press [tab] to reveal `side` [tab]. Press [🖰] to measure the length of this side of the rectangle, rather than the perimeter of the rectangle. Move the measurement using the Touchpad keys and press [🖰] to drop the length measurement in place.

3. Press [menu]⇨Measurement⇨Area to access the Area tool.

Move the cursor to any side of the rectangle and press [🖰] to measure the area of the rectangle. Move the measurement using the Touchpad keys and press [🖰] to drop the measurement in place.

Two of your measurements should be in centimeters (*cm*), the larger of which represents the perimeter. The area measurement is units of square centimeters (*cm^2*). Your diagram should resemble the first screen shown in Figure 16-2.

The next step is to lock the perimeter. This allows you to manipulate the rectangle while maintaining a fixed perimeter. To accomplish this task, move the cursor to the perimeter measurement and press ⌈ctrl⌉⌈menu⌉⇨Attributes to open the Attributes control panel. Press ▾ to highlight the Lock icon and then press ▸ ⌈enter⌉ to lock the perimeter. See the second and third screens in Figure 16-2.

Figure 16-2: Configuring the maximum area activity.

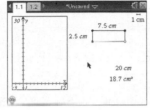 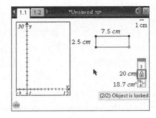 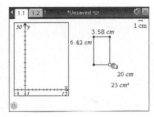

Recall that you want to analyze the relationship between the length of the rectangle and its area for a fixed perimeter. To explore this relationship, you must store these values as variables. Refer to Figure 16-3 and follow these steps:

1. **Move your cursor to the length measurement from the previous Step 2 and press ⌈⌉ to highlight the measurement (as indicated by the gray box).**

2. **Press ⌈var⌉⇨Store Var, type *length*, and press ⌈enter⌉.**

   This stores the length measurement as the variable *length.*

   A stored variable is always displayed in a bold font.

3. **Move your cursor to the area measurement and press ⌈⌉ to highlight the measurement.**

4. **Another method of storing a variable is to right-click. Press ⌈ctrl⌉ ⌈menu⌉⇨Store, type *area*, and press ⌈enter⌉.**

Figure 16-3: Storing variables.

 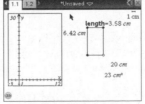 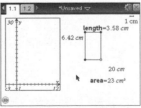

Storing a variable   Bold variable label   Storing *a* for *area*

Configuring Lists & Spreadsheet for Data Capture

Your stored variables are now available for use in any application that is part of the same problem. In the following sections, I show you how to use these variables to set up the Lists & Spreadsheet page for data capture.

Before doing so, move to the Lists & Spreadsheet page and name column A *Length_* and column B *Area_*. (The list can't be named the same thing as a variable; adding an underscore symbol to the list name avoids this problem.)

The Automatic Data Capture command collects the length and area data automatically when you move the rectangle in the Geometry application.

To configure the Automatic Data Capture command, move your cursor to the formula cell of column A and follow these steps:

1. **Press menu⇨Data⇨Data Capture⇨Automatic Data Capture.**

   This action pastes the Automatic Data Capture command in the formula cell. The cursor is positioned in the place where you input the variable name to be captured.

2. **Specify the variable by pressing var, highlighting *length*, and pressing enter.**

   Recall in the previous section that I named the length using the variable *length*. After you press enter for the last time, notice that the current length measurement is automatically placed in cell A1.

3. **Move the cursor to the formula cell for column B and repeat Steps 1 and 2. Select *area* as the variable to link to.**

   Notice that the current area measurement is placed in cell B1.

The first screen in Figure 16-4 shows the Automatic Data Capture command in the formula cell for column A. The second screen shows the command for column B.

Before moving on, move to the Lists & Spreadsheet page and name column A *Length_* and column B *Area_*. (The list can't be named the same thing as a variable; adding an underscore symbol to the list name avoids this problem.)

It is often a good idea to use an extra parameter whenever using automatic data capture to capture two variables, so that the data capture is dependent on the other variable you are capturing. The syntax for automatic data capture is `capture (variable to capture, 1, variable the capture is dependent on)`. See the third screen in Figure 16-4. This technique assures that the data is properly collected each time one of the variables changes.

Figure 16-4:
Configuring
Lists &
Spread-
sheet for
automatic
data
capture.

Notice in Figure 16-4 that the entire Data Capture command is revealed in the formula cell. The 1 following the variables length and area indicates automatic data capture. A 0 indicates manual data capture.

Graphing a Scatter Plot of Your Data

You are now ready to configure a scatter plot of the *area_* versus *length_* data. Here's what you need to do:

1. **Move back to the Geometry page and press** [tab] **to activate the entry line (as indicated by a blinking cursor located next to the first available function).**

2. **Press** [menu]⇨**Graph Type**⇨**Scatter Plot to switch the entry line to scatter plot. Alternatively, press** [ctrl] [menu]⇨**Scatter Plot while the cursor is located in the entry line.**

   In Figure 16-5, I pressed [var] to view a list of choices for the *x*-variable.

3. **Using the** ▲▼ **keys, highlight** *length_* **and press** [⊡] **or** [enter].

4. **Press** [tab] **to move to the *y*-variable, press** [var] **to view a list of choices for the *y*-variable, highlight** *area_*, **and press** [⊡] **or** [enter].

You may need to adjust your window settings (press [menu]⇨Window⇨ Window Settings) to view the captured data.

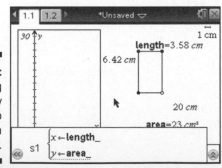

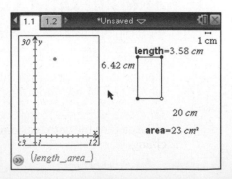

Figure 16-5:
Configuring
a Geometry
page to
display a
scatter plot.

Collecting Data Automatically

To capture data automatically, just drag the object that changes the values of your stored variables. In this activity, manipulating the rectangle changes both the length and area measurements. Consequently, these updated measurements are periodically dropped into the Lists & Spreadsheet application and plotted in the coordinate plane at the same time.

Recall that when you constructed the rectangle, you defined three vertices. You can drag the third vertex of the rectangle (or the fourth vertex that was never defined) to change the size of the rectangle. I like to change the attributes of the movable points to Empty Large. Remember that the perimeter has been locked, which adds the constraint that the perimeter must remain fixed at all times.

Move the cursor to an appropriate vertex of the rectangle and press [ctrl][🗊] to grab it. Resize the rectangle using the arrow keys on the Touchpad. As shown in the first two screens in Figure 16-6, a scatter plot of the *area* versus *length* data automatically appears. The third screen in Figure 16-6 shows the corresponding numerical data on the Lists & Spreadsheet page.

Figure 16-6:
Results of the automatic data capture activity.

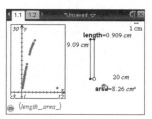

You may be tempted to drag the rectangle back and forth many times. Remember, each time you manipulate the rectangle, you are adding more points to the lists. Too much of a good thing is possible, at least when it comes to data.

Manual Data Capture

With the Manual Data Capture feature, data is only captured each time you press [ctrl][.] from within any application in the problem. Manual data capture is helpful if you want to limit the amount of data that is captured. I like using manual data capture with this activity to get students to think about which pieces of data are important to collect.

The steps for configuring TI-Nspire for manual data capture are identical to those described in the preceding section, with one small exception. Press [menu]⇨Data⇨Data Capture and select Manual Data Capture rather than Automatic Data Capture.

If you configured the Lists & Spreadsheet page for manual data capture, move to the Geometry page and grab and drag the rectangle. Periodically, press [ctrl][.] to capture another data point. Figure 16-7 shows a typical sequence of results from using the Manual Data Capture feature. You can capture as many points as you want. Just make sure that you capture enough to reveal the general shape of the graph as well as the location of the length value that produces a maximum area.

Figure 16-7:
Results of
the manual
data cap-
ture activity.

 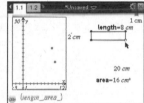 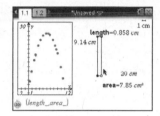

Repeating a data capture

Sometimes you want to repeat an experiment. To do this, just move to the Lists & Spreadsheet page and highlight the formula cell for column A. Press [enter] twice to clear the data in the first column while preserving the Data Capture command that you spent so much time configuring. Repeat this procedure to clear the data in column B. Now you are ready to go to the Geometry page and repeat the experiment.

Analyzing Captured Data

As a result of this activity, you now have both graphical and numerical results that can be analyzed. Here are two possibilities that come to mind:

> ✔ **Find and graph an equation to model the scatter plot.** An inspection of the scatter plot suggests that a quadratic equation can be used to model the data. In fact, if you let x represent the length of the parabola, you can find an expression for the area in terms of x, knowing that the perimeter has a fixed value.

In my example, I have a fixed perimeter of 20 centimeters. Therefore, the width of the rectangle must be $10 - x$. The formula for the area is the product of length and width, or $area(x) = x(10 - x)$. Type this equation into $f1(x)$ on the entry line to verify that it goes through the scatter plot. Don't forget to include a multiplication sign between the length and the width! See the first screen in Figure 16-8.

✔ **Sort the data to find the length that produces a maximum area.** For those of you who are more into the numbers, move to the Lists & Spreadsheet application, highlight columns A and B, and perform a descending sort by column B (see Chapter 14 for details on how to sort data). The second screen in Figure 16-8 shows how to configure the Sort dialog box. The third screen in Figure 16-8 shows that, for my example, the area is maximized for a length of approximately 4.833 centimeters.

Figure 16-8:
Analyzing
captured
data.

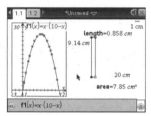

Formulas contained in the formula row are deleted anytime you do a sort on that column. Press ctrl esc to undo the sort and restore the formulas.

Part VI

The Data & Statistics and Vernier DataQuest Applications

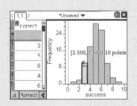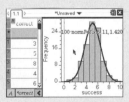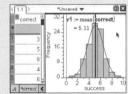

In this part . . .

This part takes you through a tour of the Data & Statistics application. I show you how this application is used to produce graphs of one- and two-variable data sets that reside in the Lists & Spreadsheet application. You also learn about the many tools available within this application to assist you with the analysis of these graphs. And, I show you how the two-way communication between the Lists & Spreadsheet and Data & Statistics applications adds a dynamic element to these analyses. Then, I show you how color can be used in this environment in the comparison of data sets.

This part also contains instructions for the Vernier DataQuest application, including an overview and instructions for conducting a data collection experiment using a CBR2 motion detector. I show you how to select the region of the data you want and display it using the three views that provide multiple representations of the data. I also show you how to replay (and repeat) an experiment. Then, I show you to set up and customize an experiment to fit your needs. I also discuss the built-in Motion Match feature.

Chapter 17

Constructing Statistical Graphs

. . . .

In This Chapter

▶ Creating single-variable and two-variable Quick Graphs from the Lists & Spreadsheet app

▶ Understanding different page layout options

. . . .

*T*he Data & Statistics application operates in conjunction with the Lists & Spreadsheet application. It provides a visual representation of numeric or categorical data using a nice variety of common single-variable and two-variable statistical graphs.

You can also use the Data & Statistics application in conjunction with the Calculator or Notes application, although these methods are less common.

In this chapter, I show you how to launch the Data & Statistics application directly from the Lists & Spreadsheet application using the Quick Graph feature. I also show you an alternative method of producing graphs.

Launching Data & Statistics from the Lists & Spreadsheet Application

As with any TI-Nspire application, you have the option of launching the Data & Statistics application by inserting a new page and selecting Data & Statistics from the seven available choices. However, because the purpose of the Data & Statistics application is to graph numerical (and sometimes categorical) data, it makes sense to start with the Lists & Spreadsheet application. Then, use the Quick Graph feature in the Lists & Spreadsheet application to launch the Data & Statistics application.

Entering data in Lists & Spreadsheet

The rules for using the Lists & Spreadsheet application in conjunction with Data & Statistics are quite straightforward. You must have one or two named lists in the Lists & Spreadsheet application.

For example, look at the lists displayed in Figure 17-1. In the first screen, I've created a list called *dice_sum* and used the command =randint(1,6,50) + randint(1,6,50) to produce a list of 50 numbers. This list represents a simulation of rolling two dice 50 times and recording their sums. I call this a single-variable data set because only one list of data is to be analyzed.

The second screen shows the *U.S. Immigrant Population* versus *Year* data that I use in Chapter 15, which represents a two-variable data set with the independent variable represented by *Year* and the dependent variable represented by *U.S. Immigrant Population.*

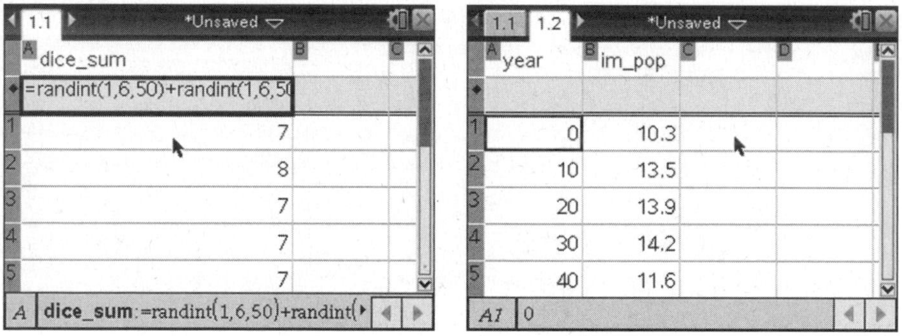

Figure 17-1:
Setting
up Lists &
Spread-
sheet for
use with
Data &
Statistics.

1.1 ▶		*Unsaved ▽	
A dice_sum		B	C
=randint(1,6,50)+randint(1,6,50			
1	7		
2	8		
3	7		
4	7		
5	7		
A **dice_sum**:=randint(1,6,50)+randint(▸			

1.1 1.2 ▶		*Unsaved ▽		
A year	B im_pop	C	D	
1	0	10.3		
2	10	13.5		
3	20	13.9		
4	30	14.2		
5	40	11.6		
A1	0			

Notice that I have named each list. You must have named lists to use the Data & Statistics application.

Using Quick Graph

After your named lists are complete, press menu⇨Data⇨Quick Graph (or right-click) to open the Quick Graph feature. The first screen in Figure 17-2 shows the result of using Quick Graph with the single-variable data set. TI-Nspire automatically opens a split screen with a Data & Statistics page side by side with the existing Lists & Spreadsheet page. By default, single-variable data sets are visually represented by a dot plot. TI-Nspire can represent single-variable data sets with a dot plot, box plot, or histogram.

The second screen in Figure 17-2 shows the result of using Quick Graph with the two-variable data set. Again, TI-Nspire automatically opens a split screen with a Data & Statistics page and the existing Lists & Spreadsheet page.

Figure 17-2:
Using Quick
Graph with
single-
variable and
two-variable
data sets.

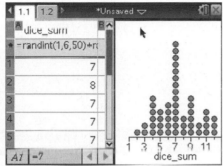
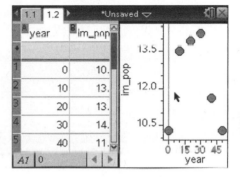

With the Lists & Spreadsheet page active, try pressing [ctrl][R] to recalculate the values dependent on the formula =randint(1,6,50) + randint(1,6,50). This action generates a new set of values, and the dot plot updates accordingly.

If you have two or more columns of data contained in a Lists & Spreadsheet application, you must select the column or columns of data to be used with Quick Graph. Here's how you accomplish this task:

✔ **Select a column.** Use the Touchpad to move the cursor to the letter at the top of the column. Click (🔲) directly on the letter. For example, click the A to select column A. This is just one of many "clickable areas" on the TI-Nspire.

✔ **Select multiple rows or columns.** After selecting a column, press and hold the [⇧shift] key. While holding this key, press the Touchpad keys to select additional columns.

Working with Data & Statistics on a Separate Page

As you see in the previous section, Quick Graph reconfigures your page layout to display a Data & Statistics graph on the current screen (unless your current page layout contains four applications).

To view a Data & Statistics graph on a separate page, press 🏠on⇨Data & Statistics (or press ⌃🄸 and select Add Data & Statistics from the available choices). The first screen in Figure 17-3 shows the result of this action. By default, the caption located at the top of the screen shows the first available list. Move your cursor to the Click to Add Variable region, press 🔳 to view the available lists (as defined in all Lists & Spreadsheet applications within the current problem), highlight your list choice, and press enter.

The second and third screens in Figure 17-3 show the result of selecting the list called *dice_sum*.

Figure 17-3:
Graphing
a single-
variable
data set on
a separate
Data &
Statistics
page.

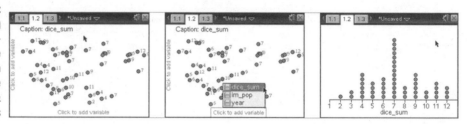

Similarly, you can add a separate Data & Statistics page to view the scatter plot corresponding to a two-variable data set.

After you select your variable on the horizontal axis, you will not be able to "see" the location to add the variable on the vertical axis. Move your cursor to the middle of the very left side of the screen and the `Click to add variable` message magically appears. Alternatively, press tab to locate the add variable box on the vertical axis.

As can be seen in Figure 17-4, you must select the variable to be represented on both the horizontal and vertical axes.

Figure 17-4:
Graphing a
two-variable
data set on
a separate
Data &
Statistics
page.

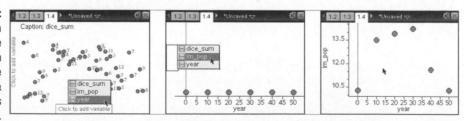

Chapter 18

Working with Single-Variable Data

. . . .

In This Chapter

▶ Constructing three types of single-variable statistical graphs

▶ Using a variety of tools to customize your graphs

▶ Creating graphs of categorical data

▶ Finding out how to manipulate single-variable data sets and graphs

. . . .

*I*n this chapter, I show you the choices of graphs available for use with numerical and categorical data contained in a single list and describe the tools that are available for use with each of these graphs. I also show you how to manipulate either the data values or the graphs themselves and observe the corresponding effects of such actions.

Selecting a Statistical Graph

Three types of graphs are available when you work with a single list of numerical data: dot plot, box plot, and histogram. When working with categorical data, you have the option of creating a dot chart, bar chart, or pie chart.

To select a graph type, press [menu]⇨Plot Type and select the plot type from the list of available options.

Dot Plots

In the first screen in Figure 18-1, I give a variation of the two-dice sum example from Chapter 17. The first column (called *first_die*) uses the command =randint(1,6,50) to simulate rolling one die 50 times. The second column (labeled *second_die*) represents the outcome of rolling a second die 50 times, and the third column (labeled *total*) represents the sum of the first two columns.

The second screen shows the result of pressing [menu]⇨Data⇨Quick Graph. The current page layout is configured with a Data & Statistics page, and the default dot plot of the data contained in the *Total* column is graphed automatically.

Changing variables

Perhaps you want to view a dot plot depicting the distribution of the *first_die* data. To accomplish this task, follow these steps:

1. **Move the cursor to the horizontal axis label at the bottom of the screen until the words** `Click or Enter to change variable` **appear.**

2. **Press 🔲 to reveal a list of choices.**

3. **Use the Touchpad keys to highlight your choice and press [enter].**

The third screen in Figure 18-1 shows the result of changing the horizontal axis category from *total* to *first_die*. Notice that the scale on the horizontal axis changes automatically. I also have the option of changing this category to display the *second_die* data.

Figure 18-1:
Changing
variables on
a Data &
Statistics
graph.

 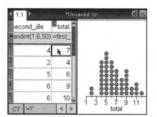 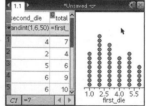

If you press [menu]⇨Plot Properties, you see menu options to add or remove X and Y variables. These options enable you to switch back and forth between one- and two-variable data sets. They also allow you to switch your plots from a vertical orientation to a horizontal orientation and vice versa.

Consider, for example, the vertically oriented dot plot shown in the second screen in Figure 18-1. You can press [menu]⇨Plot Properties⇨Remove X Variable to effectively remove the graph from the screen. Next, move the cursor to the left side of the screen until the words `Click or Enter to add variable` appear, press [⬚], and select *total* from the list. You see the same dot plot as before, but with a horizontal orientation. Alternatively, select *total* as the Y variable (instead of the X variable) and watch the orientation change.

Changing your window settings

The first screen in Figure 18-2 shows the results of a two-dice sum experiment in which a sum of 2 and a sum of 12 never occurred. TI-Nspire, therefore, produces a horizontal scale that ranges from 3 to 11. To change these settings to range from 2 to 12, press [menu]⇨Window/Zoom⇨Window Settings and change the XMin to 1.5 and XMax to 12.5. Alternatively, you can grab ([ctrl][⬚]) the tick marks and use the Touchpad to stretch or shrink the horizontal scale to make the desired adjustments. See the second screen in Figure 18-2.

In the third screen, the *x*-axis variable is switched back to *first_die*. I then pressed [menu]⇨Plot Properties⇨Force Categorical X, which treats each value in the list as categories, much as it would if this list contained words. Forcing a categorical variable is a nice option that gives a clean-looking graph and is especially well-suited for a discrete variable such as the numerical value on a die. Switching to categorical data also gives the option of pressing [menu]⇨Plot Type and selecting Bar Chart or Pie Chart.

When numerical data are switched to categorical data, TI-Nspire sorts by the first digit it sees. If you switched the dot plot for *total* from numerical to categorical, the values on the horizontal axis would go 10, 11, 12, 2, 3, . . . , 9 from left to right. Fortunately, you can grab the labels and move the data columns so that the totals are in numeric order (or whatever order you wish).

Figure 18-2:
Changing window settings and forcing categorical data.

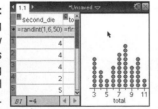 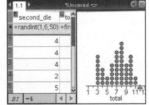 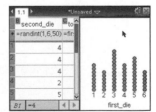

Press [menu]⇨Plot Properties⇨Force Numerical X to switch back to numeric values.

Histograms

Press menu⇨Plot Type⇨Histogram (or right-click near the middle of the screen and choose Histogram) to display single-variable data as a histogram. The first screen in Figure 18-3 shows a histogram of the *total* data.

The number of *bins,* the name given for the bars in a histogram, is determined by the number of data sets and the distribution of the data. As shown in the first screen in Figure 18-3, you can click a bin to display the width of the bin and the number of values contained in the bin.

Changing the scale

Press menu⇨Plot Properties⇨Histogram Properties⇨Histogram Scale and select a scale for your histogram, as follows:

- ✔ **Frequency:** By default, histograms give the *frequency* of each bin. This scale tells you how many values are contained in each bin. Referring to the first screen in Figure 18-3, eight values are contained in the bin ranging from 6.5 to7.5.

- ✔ **Percent:** Press menu⇨Plot Properties⇨Histogram Properties⇨Histogram Scale⇨Percent to change the scale to *percent*. This scale tells you the percent of all values that are contained in each bin. The second screen in Figure 18-3 shows that 16 percent of all values fall in the bin from 6.5 to 7.5, which makes sense because the ratio 8 out of 50 is equal to 16 percent.

- ✔ **Density:** Press menu⇨Plot Properties⇨Histogram Properties⇨Histogram Scale⇨Density to change the scale to *density*. (See the third screen in Figure 18-3.) The density is calculated by dividing the relative frequency of a bin by the bin width.

Figure 18-3: Plotting histograms and adjusting the scale.

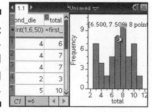 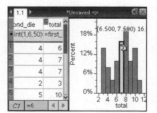 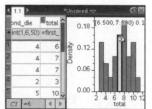

Adjusting the bins

To adjust the bin width, follow these steps:

1. **Move the cursor to the right edge of any bin until the ⊹ symbol appears.**

2. **Press** ⌈ctrl⌉⌈✑⌉ **to grab the bin and use the Touchpad keys to adjust the width.**

   Notice that the numerical value associated with right edge of the bin is displayed.

3. **Press** ⌈esc⌉ **when finished.**

To adjust the bin width numerically, follow these steps:

1. **Press** ⌈menu⌉⇨**Plot Properties**⇨**Histogram Properties**⇨**Bin Settings (or right-click on a bar and choose Bin Settings).**

2. **Choose Equal Bin Width or Variable Bin Width.**

   Choosing Equal Bin Width prompts you to input the width and the alignment. The bin width is fairly self-explanatory — it sets the width of each bin of your histogram. The alignment is a bit trickier. For example, if you specify a bin width of 2 with an alignment of 5, you can be assured that one of your bins has a right edge located at 5 with a width of 2. All other bins fall into place. A bin width of 2 and an alignment of 7 (or any odd integer) produces the same histogram.

   Perhaps you want your bins to have a width of 1 and to be centered on the values represented by the possible two-dice sum outcomes. Use a bin width of 1 and any bin alignment with a decimal portion equal to 0.5. For example, a bin alignment of 7.5 works.

   Choosing Variable Bin Width allows you to choose the name of a list where you have defined the classes. Make sure that all data points belong to a class. For example, if your list values ranged from one to five, you could have this list of bin values: {0.5, 3.5, 4.5, 5.5}.

3. **Click OK when finished.**

Changing variables

As with the dot plot, you can change variables or use the Add/Remove Variable options located on the ⌈menu⌉⇨Plot Properties submenu. Remember to use the Add/Remove tools to change your histogram from a vertical orientation to a horizontal orientation and vice versa.

Inserting a normal curve

The command =sum(randint(0,1,10) can be used to simulate guessing on 10 straight true/false questions. This formula generates a list of 10 integers that are either 0 or 1 and calculates the sum. If you assign a 1 to represent a correct answer, a result of 6 means that you got 6 out of 10 questions correct.

First, I used the Fill command to copy this formula to cell A100. I then used Quick Graph to graph a histogram of the data as shown in the first screen in Figure 18-4.

This binomial experiment should have a shape that resembles a normal distribution. Press [menu]⇨Analyze⇨Show Normal PDF to overlay a normal curve on your histogram. The normal graph is based on the mean of the data set (in this case, the mean is 5.01) and the standard deviation (1.648). See the second screen in Figure 18-4.

Plotting a value

I can also plot a value on a histogram that displays as a vertical line perpendicular to the horizontal axis at a point equal to the specified value. In the third screen in Figure 18-4, I pressed [menu]⇨Analyze⇨Plot Value. At the prompt, I typed **mean(correct)** and pressed [enter]. (See the third screen in the figure.)

You can plot a single number or an expression that equals a number. Statistical values such as mean or standard deviation are good choices for Plot Value.

The Plot Value feature is also available with dot plots and box plots.

Figure 18-4: Displaying a normal curve and plotting a value.

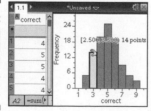 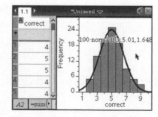 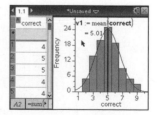

To remove a plotted value, move the cursor over the vertical line and press [ctrl][menu]⇨Remove Plotted Value. To hide a normal curve, press [menu]⇨ Analyze⇨Hide Normal PDF.

A short course in box plots

Box plots (sometimes referred to as *box-and-whisker plots*) consist of four regions. The leftmost region and the rightmost region are aptly named *whiskers.* The box located between the whiskers is divided into two regions. The dividing line in the box corresponds to the median of the data set. To draw these four regions, you need five numbers (called the five-number summary): the minimum value, the median, the maximum value, the lower quartile (the median of the lower half of the data set), and the upper quartile (the median of the upper half of the data set).

The following figure shows two ways to draw a typical box plot. The first figure does not delineate *outliers,* data values that are numerically distanced from other data values. The second figure highlights two values that are outliers. Points that are a distance 1.5 times the interquartile range beyond the quartiles are considered outliers and are drawn as distinct points. The *interquartile range* is defined as the difference between the upper quartile and the lower quartile.

Each of the four regions created by a box plot contains 25 percent of the data values. As a result, box plots are helpful in displaying the spread and skew of the data.

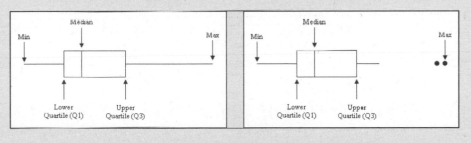

Try hovering on the normal curve, right-clicking ([ctrl][menu]), and choosing Shade Under Function. Shade a region under the curve by clicking the lower and upper bounds of the interval to display the corresponding probability associated with this interval.

Box Plots

Consider that you want to investigate the major-league home run leaders between the years of 1988 and 2002. I realize this is somewhat controversial, but from a data standpoint, the results are quite interesting!

In the first screen in Figure 18-5, I have entered the data in column A (named *al* for American League) and column B (named *nl* for National League). If you want to enter all the data, take a peek at the second and third screens to see the rest of the data in columns A and B. I then pressed [menu]⇨Data⇨Quick Graph and, by default, obtained a dot plot for the American League data.

Press [menu]⇨Plot Type⇨Box Plot to switch from a dot plot to a box plot. I still think it is cool to watch the points dynamically move and morph into the new plot every time it happens!

As you move the cursor over the box plot, each of the values that comprise the five-number summary are revealed. If you click one of the four regions in a box plot, the individual data points are revealed. Keep in mind that multiple occurrences of the same number are displayed as a single point. To hide points, move to an open space on the screen and press [🔒].

Outliers

To graph a second box plot containing the National League data, press [menu]⇨Plot Properties⇨Add X Variable and choose nl to create the second box plot.

The second screen in Figure 18-5 shows the second box plot representing the National League results. Notice that two outliers appear on this graph.

To hide outliers, press [menu]⇨Plot Properties⇨Extend Box Plot Whiskers. To reveal outliers (if they exist), press [menu]⇨Plot Properties⇨Show Box Plot Outliers.

Effective Use of Color

TI-Nspire does not automatically use color in a Data & Statistics environment. I like to use color when I am comparing data; color helps to differentiate between the different data representations. To change the AL box plot to red, press [ctrl] [menu]⇨Color⇨Fill Color. I repeated this step to change the NL box plot to blue. To further make the connection between the data in the table and the data represented in the box plot, I changed the text color in the table to match the box plots. First, select the column by clicking the capital letter at the top of each column, then right-click by pressing [ctrl] [menu]⇨Color⇨Text Color. See the results in the third screen in Figure 18-5.

Figure 18-5: Creating box plots and using color.

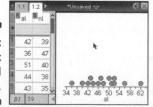 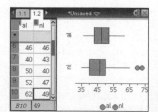 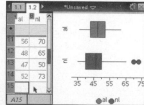

Raw Data versus Summary Data

Categorical data can be represented in two modes of statistical plotting, raw data and summary data. Raw data is contained in one list. Summary data contains separate lists. In this first example, I will use raw data to plot the favorite seasons of a group of students.

Categorical data requires the use of string values. To type a string into a cell, you can enclose the characters in quotation marks. However, in the first screen in Figure 18-6, I didn't bother to use quotes when I typed in column A, and TI-Nspire still treated the data as categorical in nature.

To switch the plot to a bar chart, press [menu]⇨Plot Type⇨Bar Chart. The second screen in Figure 18-6 shows a bar chart of the favorite seasons data. Notice that you can click a bar to reveal the number of cases and the corresponding percent for the selected category.

TIP

The data defaults to an alphabetical order (Fall, Spring, Summer, and Winter). However, you can grab ([ctrl][✋]) the labels and use your Touchpad to move the labels if you want to change the order of the display. For example, grab the word *fall* and move the data column to change the order of the chart. I think that the order: Spring, Summer, Fall, and Winter may make more sense in the given context. Additionally, you can right-click on the chart, [ctrl][menu]⇨Sort, and choose one of three options: List Order, Value Order, and Alpha Order.

Categorical case data can also be represented as a pie chart. Press [menu]⇨ Plot Type⇨Pie Chart to switch to a pie chart. The third screen in Figure 18-6 shows the pie chart of my favorite seasons data. Once again, you can click a sector to reveal the number of cases and the corresponding percent for the selected sector. To show the percentages on the pie chart, right-click, [ctrl][menu]⇨Show All Labels. Of course, if you don't want to see the percentages of each sector, right-click, [ctrl][menu]⇨Hide All Labels.

Figure 18-6: Graphing case data.

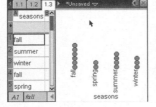 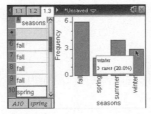 

Perhaps your categorical data is given in summary data form. That is, one column gives the category name and a second column gives the frequency of occurrence. On TI-Nspire, you can graph this data using a summary plot.

To graph a summary plot, follow these steps:

1. **Highlight the column containing the category names.**

2. **Press** [menu]⇨**Data**⇨**Summary Plot.**

3. **A dialog box appears (see the first screen in Figure 18-7).**

4. **Select the list name containing the frequency values and press** [enter] **(see the second screen in Figure 18-7).**

Figure 18-7:
Graphing
data given
in summary
data form.

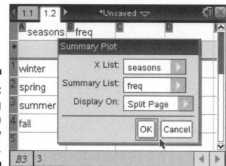 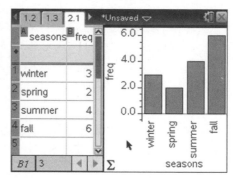

Change the categorical split data to a pie chart by pressing [menu]⇨Plot Type⇨ Pie Chart or right-click and choose Pie Chart.

Manipulating Single-Variable Data

TI-Nspire allows two-way manipulation of data, which means that you can change values in the Lists & Spreadsheet application and watch a Data & Statistics graph update automatically. Likewise, you can click and drag dot plot points, histogram bins, and box plots and observe updates in the corresponding numerical data. Plotted values, such as the mean, update automatically as well.

Manipulating dot plots

The first two screens in Figure 18-8 show the result of moving the point corresponding to a value of 52 to a location with a value of 36. Remember, to grab a point, move the cursor to the point and press [ctrl][🔒]. Then use the Touchpad keys to move the point and press [enter] to release it.

In the third screen, I selected and moved two points. To select multiple points, click each point individually. Then press ⌃ 👆 to grab and move them as you want. To deselect points, move to an open space on the screen and press 👆.

Figure 18-8:
Manipu-
lating a dot
plot.

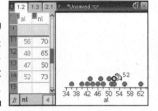  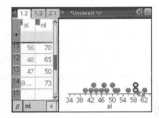

Press ⌃ esc to undo a manipulation.

Referring back to Figure 18-1, this data is generated by typing commands in the formula row (the formula row is denoted by the ♦ symbol). You cannot manipulate the graph of a data set that is dependent on a formula located in the formula row. However, you can manipulate graphs in which formulas are located directly in the cells of the spreadsheet.

If you manipulate data that is generated from formulas, the formulas are deleted and replaced with their numerical values.

Manipulating histograms

Figure 18-9 shows the result of grabbing a bin containing one value and dragging it to a bin containing two values. The bin corresponding to values 39–41 is now empty, and the bin corresponding to values 43–45 contains three values. You can be assured that the data values contained in the Lists & Spreadsheet application have changed as well.

Figure 18-9:
Manipu-
lating a
histogram.

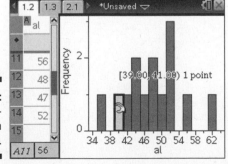 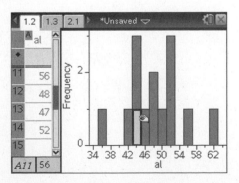

TIP

You may need to change your window settings after manipulating a histogram.

Manipulating box plots

You can grab and move any of the four quartiles contained in a box plot. The first screen in Figure 18-10 shows that I've pressed [ctrl][🖱] and grabbed the second quartile. The number 44 displayed on the screen corresponds to the lowest data value contained within this quartile.

In the second screen in Figure 18-10, I used the Touchpad keys to translate all the points contained within this quartile 47.05 – 43 = 3.05 units to the right. Press [enter] to complete the translation and release the quartile.

Figure 18-10:
Manipu-
lating a box
plot.

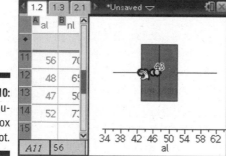 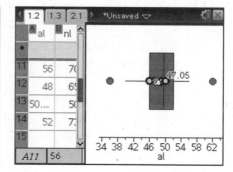

Chapter 19

Working with Two-Variable Data

In Chapter 18, I talk about single-variable data. In this chapter, I feature two-variable data sets. In addition to graphing scatter plots of two-variable data, I showcase some of the tools available in the Data & Statistics application that help you get the most out of your graphs.

I also discuss some of the pros and cons related to using the Data & Statistics application rather than the Graphs or Geometry applications to represent two-variable data.

Creating an x-y Statistical Plot

Once again, I refer to the U.S. population data contained in Table 15-1. I have opened a new document and entered the *year* and *U.S. population* data into the first two columns of a Lists & Spreadsheet page. The first column, titled *year*, represents years after 1900. The second column, titled *us_pop*, gives the corresponding U.S. population in millions.

Using Quick Graph

Highlight both columns and press menu⇨Data⇨Quick Graph or right-click (ctrl menu) and choose Quick Graph. This action automatically splits the page and adds a Data & Statistics application with the *x-y* plot of the *us_pop* versus *year* data. Notice in the first screen in Figure 19-1, I can move the cursor to any point and press ⬚ to reveal its coordinates.

Graphing on a separate page

Sometimes, it is preferable to use the Data & Statistics application on a full page. To graph this scatter plot to a separate page, follow these steps:

1. **Enter the data on a Lists & Spreadsheet page.**

2. **Press ctrl doc▾⇨Add Data & Statistics.**

3. **Move your cursor to the Click to Add Variable region at the bottom of the screen, press ⬚ to view the available lists, highlight *year,* and press enter.**

4. **Move your cursor to the left side of the screen until the** Click or Enter to add variable **message appears, press ⬚ to view the available lists, highlight *us_pop,* and press enter.**

See the second and third screens in Figure 19-1.

Figure 19-1: Creating an *x-y* scatter plot.

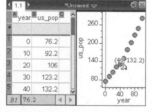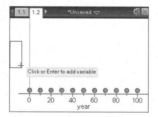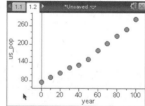

Exploring x-y Scatter Plots

In the following sections, I talk about many of the options and tools available for use with an *x-y* scatter plot.

Changing the plot type

You have two ways to view a two-variable data plot in the Data & Statistics application: as an *x-y* scatter plot or as an *x-y* line plot. The *x-y* line plot is identical to the scatter plot except that adjacent points on a *x-y* line plot are joined with a segment.

Press menu⇨Plot Type⇨XY Line Plot to switch to an *x-y* line plot. Press menu⇨ Plot Type⇨Scatter Plot to revert to an *x-y* scatter plot. Alternatively, move your cursor to the plot and right-click (ctrl menu) to choose Connect Data Points or Hide Connecting Lines.

Adding a movable line

The data in my *x-y* scatter plot looks nearly linear (although an exponential function provides a better fit). To add a movable line to a graph, press menu⇨Analyze⇨Add Movable Line.

Here are three ways to manipulate a movable line:

- **Perform a translation:** Move the cursor to the center of the line until the ✛ symbol appears. Press ctrl 🖰 to grab the line, and use the Touchpad keys to translate it. This changes the *y*-intercept of the line.

- **Perform a rotation:** Move the cursor away from the center of the line until the ↻ symbol appears. Press ctrl 🖰 to grab the line, and use the Touchpad keys to rotate it. This action changes the slope of the line.

- **Lock the *y*-intercept at 0:** Press menu⇨Analyze⇨Lock Intercept at Zero to lock the *y*-intercept at 0. If you choose to lock the *y*-intercept at 0, the movable line can only be rotated. Press menu⇨Analyze⇨Unlock Movable Line Intercept to unlock the *y*-intercept. I like this choice for direct variation problems.

Figure 19-2 shows the movable line feature in action.

Notice that the movable line is stored to the variable *m1(x)*. This variable is available for analysis anywhere within the same problem.

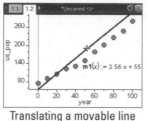 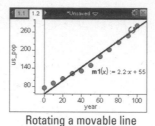 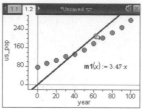

Figure 19-2:
Manipu-
lating a
movable
line.

Translating a movable line   Rotating a movable line   Rotating with a
locked intercept

Showing residual squares

Press menu⇨Analyze⇨Residuals⇨Show Residual Squares to view the residual
squares of a movable line (see Figure 19-3). Each square has sides whose
length equals the difference between the *y*-value of a given data point and the
corresponding *y*-value on the movable line.

The sum of the areas of the residual squares is also displayed on the screen.
The *least squares regression line* is the line for which the sum of the residual
squares is minimized. Grab the movable line to get as small a sum as
possible.

Try pressing menu⇨Analyze⇨Residuals⇨Show Residual Plot. As shown in the
last screen in Figure 19-3, this action produces a scatter plot of the residuals.
Don't forget to try dragging the movable line to observe the corresponding
changes to the residual plot.

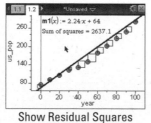 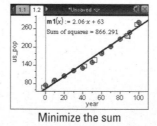 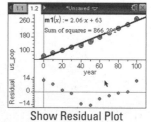

Figure 19-3:
Manipu-
lating a
movable
line.

Show Residual Squares   Minimize the sum   Show Residual Plot

Move your cursor to the movable line and press ctrl menu to bring up the context menu. The context menu gives you the choice of removing the movable
line and hiding or showing the residual squares and the residual plot.

Performing regressions

Press menu⇨Analyze⇨Regression to view a list of available regressions (the same list that is available in the Lists & Spreadsheet application). Select Show Linear ($mx + b$) or Show Linear ($a + bx$) to view the least squares regression line for the data set. Show Linear is a good feature to use after experimenting with a movable line.

The regression line associated with the U.S. population data is shown in the first screen in Figure 19-4. In the second screen I added a residual plot to the regression equation. The third screen shows that I can press menu⇨Analyze⇨Graph Trace to trace along a graph.

Figure 19-4:
Performing
a linear
regression
on the U.S.
population
data.

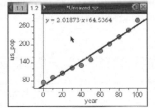 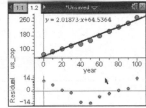 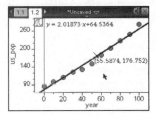

TIP

You can have multiple regressions and/or a movable line displayed on the same scatter plot. Click each graph to highlight it and to reveal its equation. Furthermore, if you choose to reveal the residual squares or the residual plot, only the residuals associated with the currently selected graph are displayed.

Plotting a value

As with single-variable data sets, I can plot a value on a scatter plot that is displayed as a vertical line perpendicular to the *x*-axis at a point equal to the specified value. To access the Plot Value feature, press menu⇨Analyze⇨Plot Value. At the prompt, type the value (or expression that yields a numerical value) and press enter to draw the vertical line associated with the value.

REMEMBER

You can plot a single number or an expression that equals a number. Statistical values such as mean or standard deviation are good choices for the Plot Value feature.

REMEMBER

To remove a plotted value, move the cursor over the vertical line and right-click (ctrl menu); then choose Remove Plotted Value (or del).

Plotting a function

As I mention earlier, the best fit for the U.S. population data is an exponential function. To graph a function, press menu⇨Analyze⇨Plot Function. Enter the function at the prompt, and press enter to view the graph.

Adding a slider

To get really fancy, try adding sliders to your graph (press menu⇨Actions⇨Insert Slider). In the first screen in Figure 19-5, I created two minimized sliders, one to manipulate the value of a in the function $y = ab^x$ and one to manipulate the value of b in the function $y = ab^x$.

In the second screen in Figure 19-5, I have opened the Plot Function tool and typed the function **a·b**^{x} at the prompt. The boldface letters indicate that I am working with the stored variables defined in my sliders. The third screen in Figure 19-5 shows my attempt to get a nice match for the data set.

Figure 19-5:
Plotting a function with a slider.

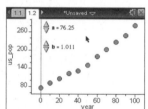 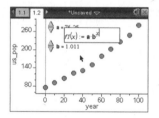 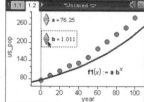

Shading under a function

Press menu⇨Analyze⇨Shade Under Function to shade under a function. Then follow these steps:

1. **Move the cursor to a desired location and press ⌨ to set the left bound of the shaded region.**

   Observe the dotted line and the number indicating the current x-value.

2. **Move the cursor to a desired location and press ⌨ to set the right bound of the shaded region.**

3. **A shaded region appears along with a number representing the area of the region.**

   Click and drag the left or right side of the shaded region to change the endpoints. See Figure 19-6.

Figure 19-6:
Finding the
area under
a curve.

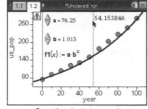
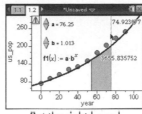
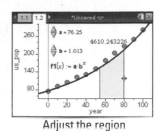

Set the left bound   Set the right bound   Adjust the region

Adjusting your window settings

Press [menu]⇨Window/Zoom⇨Window Settings to manually change the XMin,
XMax, YMin, and YMax values. Remember to press [tab] to move through each
field, and press [enter] at any time to close the dialog box.

Personally, I think TI-Nspire does a great job of coming up with a good initial
window to fit a data set. However, sometimes the manipulation of a graph
or data set results in window settings that are no longer appropriate. Press
[menu]⇨Window/Zoom⇨Data to let TI-Nspire adjust the window settings for you.

Finally, press [menu]⇨Window/Zoom⇨Zoom – In (or Zoom – Out) to access the
Zoom – In or Zoom – Out tool. Position the cursor on an area of the graph in
which you want to zoom in (or zoom out) and press [clickpad]. Continue pressing [clickpad]
to zoom in or out some more. When finished, press [esc] to exit the Zoom tool.

Move your cursor to some open space on a graph, and right-click ([ctrl] [menu]) to
choose Zoom for quick access to the Window/Zoom menu.

Try moving your cursor to the horizontal or vertical axis. Press [ctrl] [clickpad] to grab
an axis, and use the Touchpad keys to change the scale. Only the selected axis
changes.

Manipulating Two-Variable Data

As with single-variable data, you have the option of changing data values in
the Lists & Spreadsheet application and dragging points directly on a scatter plot. Keep in mind that the two-way communication established between
these two applications allows changes in one environment to be reflected in
the other.

Changing list values

If you decide to change or add a value(s) to your spreadsheet, go right ahead. Your Data & Statistics graph updates automatically. Just highlight a cell, type a new value, and press [enter] to put the change into effect.

You can also add data to the end of a list. For example, the U.S. population data from this U.S. population example goes up to the year 2000 ($x = 100$). The total U.S. population in the year 2007 was approximately 301 million. To add this data to the existing lists, simply type **107** in cell A12 (the first available cell in column A) and type **301** in cell B12. The scatter plot updates automatically. Don't forget to change your window settings to see this new point.

Dynamic regressions

To observe the real power behind manipulating a scatter plot, try moving one or more *x-y* points while the graph of a regression equation is displayed.

In the first screen in Figure 19-7, I drag a point whose current coordinates are (2.274,1.608). As this point moves, the regression equation automatically updates. That is, the graph moves and the displayed equation changes. The coordinates of this point also change in the corresponding Lists & Spreadsheet application.

To select multiple points, click each point individually. Then press [ctrl][⊞] to grab and move them as you want (as shown in the second screen in Figure 19-7). To deselect points, move to an open space on the screen and press [⊞].

In the third screen, I press [menu]⇨Analyze⇨Residuals⇨Show Plot. This gives me the option of observing changes to the residuals as I move points on the scatter plot.

Figure 19-7: Dragging points tied to a regression equation.

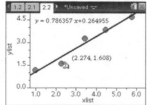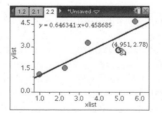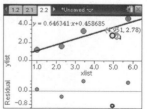

Why stop there?! Try grabbing and moving a point on the residual plot and watch the corresponding change to the scatter plot, as shown in Figure 19-8. Moving points on the residual plot adds another level of interactivity to an already dynamic graph.

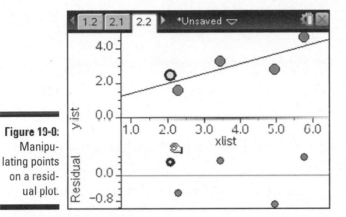

Figure 19-0:
Manipu-
lating points
on a resid-
ual plot.

Investigating Scatter Plots in Graphs versus Data & Statistics

As you read this chapter, you may wonder why you would ever use the Graphs application to graph and analyze two-variable data. Here are two reasons why you may choose to use the Graphs application over the Data & Statistics application:

- ✔ **Function graphing and analysis:** Although the Data & Statistics application offers several tools to analyze graphed functions (for example, Graph Trace and Shade Under Function), the Graphs application offers several additional options not available in Data & Statistics. For example, the Graphs application enables you to translate and stretch graphs as well as add a function table, perform calculations, construct a locus, transfer measurements, and so on.

- ✔ **Geometry tools:** The Graphs application allows you to add geometric objects including lines, segments, circles, tangents, and so on. For example, perhaps you want to add a tangent to a graph for the purpose of exploring the rate of change.

The Data & Statistics application also has several advantages over the Graphs application. In my class, we use Data & Statistics more than Graphs to graph a scatter plot. Why? Because it is easier and faster! There is no need to set the size of the window or change the graph type. I find that in most cases, the Data & Statistics application has all of the features that I need. Plus, it has some nice statistical analysis tools and I really like the way the scale is automatically labeled and formatted.

I encourage you to give these ideas some consideration. Also, with a bit of practice using both applications, you will surely gain a better understanding of which application best suits your needs.

Summary Plot with Two-Variable Data

Categorical data can be represented with a summary plot. This is one of my favorite new features on TI-Nspire. A summary plot is an excellent way to compare two data sets side by side. A group of students were asked to share their favorite food; the results can be found in the first screen in Figure 19-9. Notice that the data is separated into boy and girl categories.

After the data is typed into a Lists & Spreadsheet page, follow these steps to graph the summary plot:

1. **Highlight all three columns of data. Click the A that names the first list, press and hold the ⇧shift key, then press ▶ twice.**

2. **Perform a right-click, ctrl menu ⇨Summary Plot.**

   Fill in the Summary Plot dialog box. I only changed the *Display On* field to *New Page*. See the second screen in Figure 19-9.

3. **Press enter to plot the graph.**

4. **Right-click on a bar to change its color. Press ctrl menu ⇨Color⇨Fill Color.**

   I changed the color of the girls bars to magenta and the boys color to blue. See the result in the third screen in Figure 19-9.

You can grab and drag the labels (pizza, chips, and so on) and change the order of the display of the bars in the chart.

Figure 19-9:
Summary
plot with
two-variable
data.

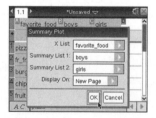 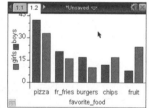

Chapter 20

Data Collection

● ●

In This Chapter

▶ Understanding the data collection process

▶ Conducting experiments automatically or manually

▶ Analyzing the data from an experiment

▶ Using the Vernier DataQuest application to customize experiments

● ●

*T*I-Nspire has the capability of collecting real-time data by attaching a variety of compatible sensors to either the handheld device or directly to a computer running TI-Nspire Computer Software. For example, you can attach a temperature sensor to your device (or computer) and collect temperature-versus-time data at a rate and duration specified by you.

In this chapter, I focus primarily on the collection of motion data (via a CBR 2 motion detector), but you can be assured that the methods and procedures that I describe for this detector can be easily adapted for use with any sensor. I also tell you about the compatible sensors that work with TI-Nspire and describe how to customize experiments to suit your specific needs.

An Overview of the Data Collection Process

The Vernier DataQuest application has many features that make it easy to collect and analyze data. Here's a brief overview of how to conduct a data collection experiment:

1. Attach a compatible sensor to a TI-Nspire device or a computer running TI-Nspire Computer Software.

   The Vernier DataQuest application automatically starts, inserting a page into your current document. See the first screen in Figure 20-1. Notice the three Infobox view tabs in the upper-left corner of the screen. Use these tabs to change the view: Meter, Graph, or Table.

2. Manually configure the Vernier DataQuest application or choose to work with the default sensor settings.

3. Run the experiment.

4. Analyze the results, which are given graphically (click the Graph View tab), numerically (click the Table View tab), or both. See the second and third screens in Figure 20-1. At the conclusion of an experiment, you have the option of saving the data and running a new experiment, replaying the experiment (change the speed as needed), or running a new experiment and overwriting the data.

Figure 20-1:
Vernier
DataQuest
Graph view
and Table
view.

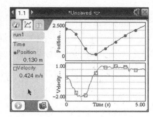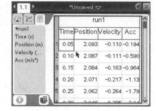

When using the Vernier DataQuest application to collect data, you should only have one document open on your Computer Software. Otherwise, the sensors will have difficulty determining which open document to use.

Compatible Sensors

At the time of this writing, TI-Nspire is compatible with more than 55 different sensors, including sensors that measure electric charge, gas pressure, light intensity, pH, sound, temperature, and voltage. Additionally, the following interface devices work with TI-Nspire:

✔ **Vernier Go!Link:** This device provides an interface between a computer and compatible Vernier sensors.

✔ **Vernier EasyLink:** This device provides an interface between the TI-Nspire Handheld and compatible Vernier sensors.

✔ **TI-Nspire Lab Cradle:** This device has three analog ports and two digital ports. Sensors can be plugged in and data can be collected from remote locations; the cradle even turns off when not sampling data. A threshold trigger can be used to start data collection. If you do a lot of data collection in your classroom, you will want to have one of these.

Conducting Experiments

Consider that you want to collect data on a bouncing ball. Your interest is in observing the successive heights of each bounce of the ball. Figure 20-2 illustrates the physical setup of this activity.

Initiating an experiment

After setting up the physical configuration in Figure 20-2, follow these steps to collect data from the experiment:

1. **Connect the CBR 2 motion detector to the TI-Nspire device using a USB cable.**

   The sensor automatically initiates the data collection process.

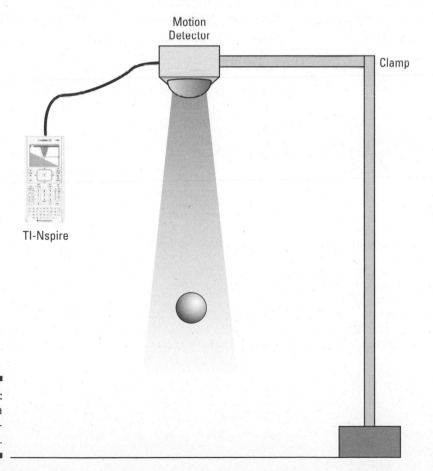

Figure 20-2:
Setting up a ball-bounc-ing activity.

2. **Configure the collection setup by pressing** menu ⇨**Experiment**⇨
   Collection Setup.

   Data is collected at a default interval and duration for the given sensor.
   The Texas Instruments CBR 2 motion detector collects data every 0.05
   seconds for 5 seconds, which corresponds to 101 data points. See the
   first screen in Figure 20-3.

3. **Press the Play button in the lower-left corner of the screen to run the**
   experiment.

   As the experiment runs, a line plot is graphed in real time, showing both
   the position and velocity graphs. The window settings in Graph view are
   automatically configured to view the data. Clicking the Table View tab
   allows you to see a display of the numeric data. Data is collected and
   displayed automatically in four lists. The variables that represent each
   list are `run1.time`, `run1.position`, `run1.velocity`, and `run1.`
   `acceleration` (see the second and third screens in Figure 20-3).

Figure 20-3:
The results
of the
bouncing-
ball activity.

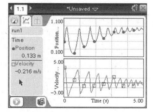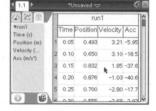

If you need to extend the duration of the experiment, press menu ⇨Experiment⇨
Extend Collection. This action extends the time to approximately 1.5 times the
current duration.

Using the Reverse feature

Because of the location of the motion detector, maximum distances are
recorded when the ball hits the ground. I used the Reverse option to give
maximum distances when the ball is *closest* to the motion detector. This
is the same effect it would have if the motion detector were on the ground
pointing up at the ball. Use the Reverse tool by pressing menu ⇨Experiment⇨
Set Up Sensors⇨Reverse. See the result in Figure 20-4.

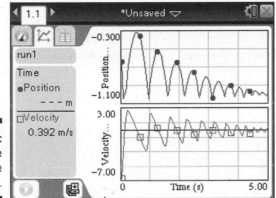

Figure 20-4:
Using the
Reverse
feature.

Replaying an experiment

I love this feature! I can see an instant replay of my experiment. What a great tool to use in the classroom! To replay an experiment, follow these steps:

1. **Press** [menu]⇨**Experiment**⇨**Replay**⇨**Playback Rate.**

   Use one of the preset playback rates, or enter your own. A useful aspect of this feature is that you can speed up slower experiments and slow down faster ones to facilitate livelier class discussions. Choosing a rate lower than 1 slows the rate, and choosing a rate higher than 1 speeds the rate. Deselect the Repeat check box if you prefer to watch the replay only once.

2. **Press** [menu]⇨**Experiment**⇨**Replay**⇨**Start Playback.**

   In the lower-left corner, you find three control buttons. Use these buttons to stop, pause, or advance the playback one frame at a time. For fun, try advancing the playback one frame at a time while in Table view!

Change the units of measurement by pressing [menu]⇨Experiment⇨Set Up Sensors. Then, choose the measurement unit that you prefer.

Being more selective with your data

Maybe you want to work with only one part of your data. For example, you can isolate one parabola from the bouncing-ball data. To get a better view of the Position graph, press [menu]⇨Graph⇨Show Graph⇨Graph 1. Follow these steps to select part of your data:

1. **Select the region you would like to work with. On the handheld, click the graph and hold for one second; then use the Touchpad keys to highlight a region on the graph. On the Computer Software, click and drag to select the region.**

   See the first screen in Figure 20-5.

2. **Right-click, ctrl menu ⇨ Strike Data ⇨ Outside Selected Region.**

   Alternatively, you could have highlighted the region of the graph that you wanted to remove (the data is not deleted permanently) and chosen to strike the data Inside Selected Region. The graph only displays the data from one bounce. Notice that some of the data in the Table view has a line through it to indicate that it has been removed. The struck-through data has not been permanently deleted, just temporarily ignored. Unlike TI-84/EasyData where the data *does get* deleted, this tool allows you to restore data that has been struck. See the second and third screens in Figure 20-5.

You can restore the data that was removed. Press menu ⇨ Data ⇨ Restore Data ⇨ All Data. The graph window automatically adjusts to accommodate the change. Don't you just love working in a dynamic environment?

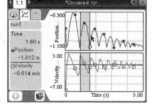

Figure 20-5: Selecting your data.

Analyzing the data

The Vernier DataQuest application has many built-in tools that allow you to analyze the data. The Infobox on the left side of the screen shows the Analyze tools that you have chosen to activate. Here I highlight a few of my favorite tools in the Analyze submenu.

Use the Statistics feature to display information about the data. Press menu ⇨ Analyze ⇨ Statistics to display your results in the Infobox on the left side of the screen, including number of samples, minimum, maximum, mean, and standard deviation. See the first screen in Figure 20-6.

The Tangent tool can be used to draw a tangent line to the curve. The slope of the tangent line is recorded in the Infobox on the left. To open this tool, press menu ⇨ Analyze ⇨ Tangent. See the second screen in Figure 20-6.

Use the Integral tool to display the area under a curve. Press menu⇨Analyze⇨ Integral to open the tool. The Graph view shades the region under the curve, and the Infobox on the left displays the calculated area under the curve.

The Analyze tools that display in the Infobox on the left side of the screen can be minimized or maximized (like a folder in My Documents). To minimize an Analyze tool, press ▼ next to the tool, or to maximize a tool, press ▶ next to the tool.

Use the Model tool to manually determine the curve of best fit for the data. Press menu⇨Analyze⇨Model to open the tool. Click the drop-down menu in the dialog box to choose the model that you want. See the third screen in Figure 20-6. After you choose the initial values for the parameters, you can adjust the parameters by clicking the ▲▼ arrows in the Infobox.

Figure 20-6:
Analyzing
the data.

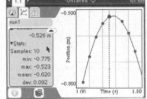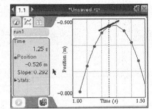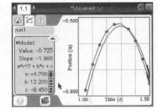

You can also automatically find the curve of best fit by performing a regression. Press menu⇨Analyze⇨Curve Fit. I chose Quadratic Regression. The graph displays in Graph view, and the function parameters are displayed in the Infobox.

The Analyze tools can be deleted by pressing menu⇨Analyze⇨Remove and choosing the tool that you would like to remove.

Repeating an experiment

To repeat an experiment, follow these steps:

1. **Store the data from your previous run by pressing the Store Latest Data Set icon (it looks like a file cabinet) near the bottom of the screen or press** menu⇨Experiment⇨Store Data Set.

You can rename the generic title, *run1,* to whatever you desire. Press the Table View tab, right-click (ctrl menu) the title, and choose *run1* Options. The Name field has *run1* highlighted. Change the name by typing the new name, and click OK to make the change take effect. See the first screen in Figure 20-7.

2. **Press the Play button (or press** menu⇨**Experiment**⇨**Start Collection) to initiate a new experiment.**

First, click the icon that looks like a file cabinet to store your data; otherwise, your data will be overwritten. If you press menu⇨Experiment⇨Start Collection, the following message appears: `This collection will overwrite the latest Data Set. Do you wish to Store or Discard the latest Data Set?` See the second screen in Figure 20-7.

Did you know that you can compare multiple data sets in the same graphing window? In the Graph view, press menu⇨Graph⇨Select Data Set⇨All (selecting More allows you to select the check box next to the runs that you would like to graph). See the third screen in Figure 20-7.

Figure 20-7:
Repeating
an experi-
ment and
displaying
multiple
runs on the
same graph.

 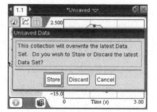 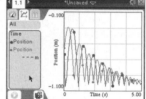

Right-click, ctrl menu⇨Color to make a change to the color scheme. Alternatively, customize the points on the graph by right-clicking and choosing Point Marker to make the adjustments you want.

Customizing Your Experiments

In the following sections, I talk about a number of ways you can configure experiments to your specifications.

Exploring the Options submenu

Of all the applications on TI-Nspire, the Vernier DataQuest application has the greatest capacity to be customized. The Options submenu provides more options and settings that can be tweaked to fit your needs, including the following:

✔ **Point Options:** The default point option is to have regional points that are connected. Regional points are spaced evenly throughout the data. You have the option of not displaying any points or displaying all the collected data points. Deselect the check box in the dialog box if you don't want the data points to be connected.

✔ **Autoscale Settings:** During the data collection, Autoscale can be turned off or set to Autoscale Larger (automatically fits the data in the screen). After the data collection has finished, you have three options: Autoscale Turned Off, Autoscale to Data (the default setting), or Autoscale from Zero (includes the origin on the graph).

✔ **Derivative Settings:** Enter the number of points (from a drop-down menu) to calculate the derivative when calculating velocity or acceleration.

✔ **Print All Settings:** Using TI-Nspire Computer Software, the screen can be printed by selecting Print Current View. Print All Views prints the Meter, Graph, and Table views. Choosing More allows you to print any combination of the three views. See the first screen in Figure 20-8.

✔ **Show/Hide Meters:** Selecting the check boxes can add additional meters (Time, Position, Velocity, and Acceleration) when displaying Meter view. See the second screen in Figure 20-8.

✔ **Hide Details:** If you want more room to display the results of your experiment, you can hide the Infobox. This option changes to Show Details after selecting Hide Details. See the third screen in Figure 20-8.

Figure 20-8:
The Options
submenu.

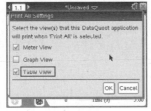  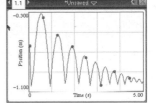

Collection mode options

You have many different ways to collect data. Press menu⇨Experiment⇨ Collection Mode to access this submenu. Here is a brief overview of the options that are available to you:

✔ **Time Based:** The default setting, this captures data with respect to time.

✔ **Events with Entry:** You can manually define a set of events when data should be collected.

- ✔ **Selected Events:** Each time that Add Data Marker is selected (the icon that looks like a camera located near the play button) from the Graph View options, data samples are manually collected.
- ✔ **Photogate Timing:** For use with a photogate sensor.
- ✔ **Drop Counting:** For use with a drop counter sensor.

Configuring the collection setup

You can access a really neat feature (with a funny name) that I don't want you to miss. If you haven't tried the Strip Chart tool (I told you it has a funny name), I promise that you will like it! Press [menu]⇨Experiment⇨Collection Setup. To use the Strip Chart tool, just select the check box in the dialog box. Make note of the time in the Duration field.

When the Strip Chart tool is activated, the sensor will continue to collect data until you press the Stop button. However, it only keeps the data for the most recent duration that you specified. Previous data is removed.

This tool gives you the freedom to have "do-overs" without bothering to start and stop the experiment. If you specify five seconds for the duration, the actual data collection may take five minutes, but only the last five seconds of data are kept intact and recorded. It is fun to watch the Graph view as the data cycles through the specified duration.

Working with the Data Control Console

If you collected data using an older OS on TI-Nspire, you may be familiar with the bar at the bottom of the screen called the Data Control Console. The Vernier DataQuest application has far more options (I think it is the better choice). However, if you want to collect data and see it displayed directly on a graph, Data & Statistics page, or Lists & Spreadsheet page, you need to at least be aware that this is still an option.

First, insert the desired application that you would like to display your collected data. Press [ctrl][D] to open the Data Control Console. Press the Play button to start collecting data.

Matching Graphs with Motion Match

Do you remember the Match the Graph feature on the TI-84 calculator? New and improved, the Motion Match feature is quick and easy to use. First, run an experiment (which sets the window for your activity). Press menu⇨Analyze⇨Motion Match⇨New Position Match to open the tool.

Press Play and try to match the graph that was randomly generated. It is not as easy as you think. See my weak attempt at matching the graph in the first screen in Figure 20-9. Lots of good math discussion and class participation arise when you use this tool in a classroom setting. I enjoy drawing the prediction so that I can customize the graph that students have to try and match. Press menu⇨Analyze⇨Draw Prediction⇨Draw to access this feature. (See the second and third screens in Figure 20-9.)

Figure 20-9:
Using
Motion
Match.

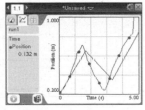 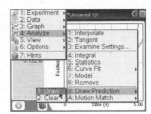 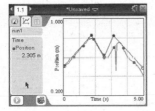

Part VII
The Notes Application

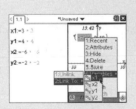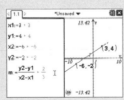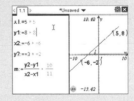

In this part . . .

This part covers the last of TI-Nspire's core applications. I show you how the Notes application can be used to add continuity to your documents and virtually eliminate the need to include separate paper instructions or questions with your TI-Nspire documents. You see that the TI-Nspire Notes application really does complete the document structure.

I also discuss how the Default, Q & A, and Proof templates contained in the Notes application can take your documents to a whole new level. In addition, I introduce you to Math Expression Boxes which can link interactively with the other applications.

Chapter 21

The Why and How of Using Notes

In This Chapter

▶ Understanding how the Notes application can enhance your documents

▶ Using the three available Notes templates for greater flexibility

*I*n this chapter, I introduce the Notes application. I talk about how Notes comprises an integral component of an entire document, and I give you some suggestions as to how you can use Notes to customize activities to suit your exact needs.

Using Notes to Complete the Document Model

Mathematical concepts can be represented in multiple ways, and TI-Nspire does a great job of accommodating these different representations. The Calculator application is well-suited for *algebraic* representations. The Graphs, Geometry, and Data & Statistics applications provide *graphical* and *geometric* representations. The Lists & Spreadsheet application specializes in *numeric* representations.

Finally, the Notes application allows the *written* or *verbal* representation of mathematical concepts. Taken together, the seven core TI-Nspire applications can be used to create dynamically linked documents that allow users to see math in multiple ways. In the next chapter, you see how math expression boxes can become dynamic, linking interactively with the other applications.

In many ways, the Notes application is the perfect complement to the other six TI-Nspire applications. For example, if you are a teacher writing an activity for use by students, the Notes application can be used to interject instructions, which eliminates the need for paper notes to accompany activities. Plus, going paperless is better for the environment! The Notes application also provides a place for teachers to pose questions and for students to type their responses.

Here is a summary of three key ways to use the Notes application:

✔ As the name suggests, the Notes application is a place to interject notes within a document. It can be used to give instructions such as "Advance to the next page and graph a function that. . . ." Notes pages provide a great way to enhance the continuity of a document and also eliminate the need for instructions on paper.

✔ The Notes application provides a place to pose and answer questions. In a classroom setting, students can type their responses directly into the Notes application. At the conclusion of an activity, students can save their work and submit it to the teacher. Now the teacher has complete electronic documentation of student work (especially with the help of the TI-Nspire Navigator).

✔ Sometimes, teachers use the Notes application in conjunction with a paper worksheet. Students may follow instructions or prompts contained within the Notes application and use a worksheet to provide paper documentation of their thoughts and ideas as well as answers to specific questions. Using Notes in conjunction with a student worksheet is a good option for those educators who prefer to keep student work and assessments in paper form.

As you see in the next sections, the Notes application comes with the tools to allow users to customize their documents for a variety of purposes.

Finding Out Which Template Is for You

The Notes application includes three templates from which to choose. In the following sections, I describe each of these templates and give you some reasons why each template is used.

The Default template

From within an existing document, press [ctrl] [doc▾]⇨Add Notes to open a new Notes page. Alternatively, press [ctrl] [I] and select Add Notes from the list of available options.

The Notes application opens in the Default template, as shown in the first screen in Figure 21-1. The Default template is the most commonly used template; it resembles a blank sheet of paper.

The second screen in Figure 21-1 shows how the blank page is typically used to provide instructions and enhance the continuity of a document. The third screen is included for reference.

Figure 21-1:
Using the
Default
template
to build a
complete
document.

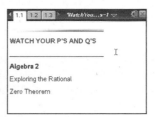 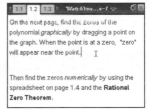 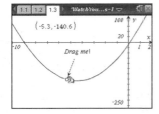

Notice the bold font in the second screen in Figure 21-1. I talk about how to customize fonts in Chapter 22.

The Q&A template

Press menu⇨Template⇨Q&A to access the Q&A template. As this name implies, this template is used to pose and answer questions.

The first screen in Figure 21-2 shows the Q&A template with a question contained in the top portion of the screen. To move the cursor to the answer portion of the screen, press tab three times. If no text is in the question area, you only need to press tab twice. You can then type your response, as shown in the second screen in Figure 21-2.

Notice the symbol located near the middle of the screen. Press tab until you move from the question region and activate this control. Press enter to collapse the answer region, which is a good option if you want to include the answer but keep it hidden from view. This icon changes to when the answer region is hidden from view. Activate the Expand/Collapse control and press enter again to expand the answer region.

Figure 21-2:
The Q&A
template.

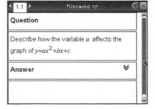 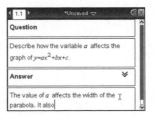 

The Proof template

Press [menu]⇨Template⇨Proof to access the Proof template.

In the first screen in Figure 21-3, I configured a page with the Notes application on the left (Default view) and the Geometry application on the right. This serves to present the problem, with a related sketch, to be proved.

In the second screen in Figure 21-3, I have opened a Notes page and configured it for the Proof template. As you can see, I've included some statements and reasons. I've asked a student to fill in the missing statements and reasons.

The third screen in Figure 21-3 contains the complete proof.

Figure 21-3:
The Proof
template.

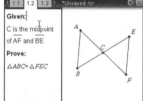

It's okay for the length of a proof to extend beyond the viewable region. If so, a scroll bar appears on the right side of the screen.

As you can see in Figure 21-3, I've included several symbols with my proof. See Chapter 22 to find out more about changing fonts and inserting special symbols.

Chapter 22

Taking Notes to a Whole New Level

In this chapter, I show you how to work with special text styles and characters. You also find out about some other tools available in the Notes application, including a tool that allows you to evaluate expressions, as you can do in the Calculator application.

Formatting Text

A range of fonts, font sizes, text styles and symbols can be added to your Notes pages. Many symbols that are used in Notes can be found in the Symbol palette (ctrl ⌨).

Selecting text

Perhaps you have typed some text and you want to go back and change a word or phrase to add emphasis. Here are the steps to follow to select a section of text to reformat:

1. **Move the cursor to the beginning or end of the word or phrase you want to change.**

2. **Press and hold the ⇧shift key.**

3. **Use the Touchpad to highlight over the entire word or phrase.**

Choosing a text format

To change the font of selected text, press menu⇨Format and select the desired text style, font, and font size. The first screen in Figure 22-1 shows the different text styles available in the Notes application. TI-Nspire Sans is the default font, but you have the option of using the more formal TI-Nspire font. Font sizes range from 7 to 24. Font styles include bold, italic, underline, superscript, subscript, and strikethrough. Unlike previous operating systems, text can be more than one style at the same time. Alternatively, after you have highlighted text, right-click (ctrl menu) to access the Format menu. The second screen shows a few time-saving options, including Cut, Copy, and Paste.

You can change text format as you type. Just press menu⇨Format and select a text style. Text that you type after the current cursor location appears in this new style.

Figure 22-1: Working with different text formats and adding color.

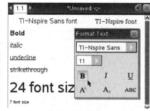

You may have noticed that no menu option exists to revert to the standard font. Instead, you turn off the current style you are working with. Consider, for example, that you have just finished typing a word in **bold** and want to switch back to standard font. Press menu⇨Format⇨Bold to turn off the bold font and change back to standard font.

Choosing a color

You can change the color of the text, or highlight the text using any one of 16 different colors. To colorize your text, you first need to highlight the text. Then, right-click, ctrl menu⇨Color⇨Text color (or Fill Color). Alternatively, you can press doc▾⇨Edit⇨Color⇨Text color (or Fill Color). If you use the latter method, the text that you type will show up in the color you selected. See the third screen in Figure 22-1.

Inserting special characters

A variety of geometric shape symbols are located within the Notes application. These shapes are particularly helpful when working with the Proof template.

To access these geometric shapes, press menu⇨Insert⇨Shape and choose the appropriate symbol. The first screen in Figure 22-2 shows each of the symbols available on the Shapes menu.

Each time you access these shapes, a small dashed box appears around the shape. Any text that you type immediately after invoking a shape is contained in this box with the symbol. This is particularly helpful when using the line, segment, ray, and vector shapes because your text appears underneath each of these symbols. Press the ▶ key or the spacebar to move out of the dashed box and resume normal typing.

A multitude of other mathematical symbols are available via the Symbol palette. As shown in the second screen in Figure 22-2, press ctrl 🔡 to access this palette, scroll through and highlight your choice, and press enter to paste it into the Notes application.

Figure 22-2: Using the Shapes menu and the Symbol palette.

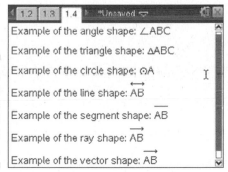
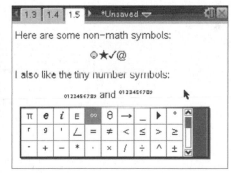

Evaluating Expressions Using Interactive Math Boxes

Press menu⇨Insert⇨Math Box to type a mathematical expression in the Notes application. Alternatively, press the shortcut key sequence ctrl M to access the Math Box option.

Pressing some keys, such as ⌨, automatically opens a math expression box.

When you open the Math Box tool, a dashed box opens and you can type your expression within this box. Press ⏎ to evaluate the math box. Press the ▸ key to move out of the dashed box and resume normal typing.

TIP

Text contained in a math expression box is italicized, just as in the Calculator application. However, some text (such as *delvar*) appears nonitalicized, indicating that you have typed a system command.

In the first screen in Figure 22-3, I typed a fraction in a math expression box and pressed ⏎ to evaluate it. Notice that the answer is in simplified form and the answer is in a different color. In the next section, I show you how to change the attributes of the math expression box so that the symbol between the expression and the answer will be an equal sign (my preference). In the next screen, I typed an expression to represent the golden ratio. Because this is not a TI-Nspire CAS, the answer displays as an approximate decimal. The third screen shows the math boxes after the attributes have been changed.

Figure 22-3:
Using the
Math Box
tool.

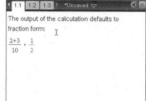 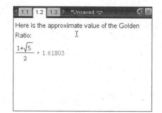 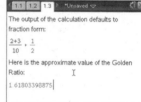

Changing the attributes of a math box

Math boxes can be customized by accessing the Attributes menu.

Here are the steps that I followed to open the Attributes menu:

1. **Type the expression using the Math Box tool (⎈ M).**

2. **Press ⌨➪Math Box Options➪Math Box Attributes, or save time by right-clicking, ⎈ ⌨➪Math Box Attributes.**

The first screen in Figure 22-4 shows the many options available for displaying the input and/or output of the math box. The second screen shows the different symbols that can be placed between the input and output of the math box. Each math box can have different settings for the display digits and the type of angle (Degree, Radian, or Gradian). In the last screen, I use a TI-Nspire CAS to show examples of a wrapped expression that displays a warning message.

Figure 22-4:
Changing
the attri-
butes of a
math box.

Did you know that you can change the color of the input or output of a math box? To change the input color, position your cursor somewhere in the input part of the math box. Press doc▾⇨Edit⇨Color to access the color tools. Place your cursor in the output part of the expression and repeat the process to change the color of the output.

Interacting with other applications

The real beauty of math boxes is their ability to seamlessly interact with other applications. In this investigation, I am going to use a Notes page to interact with a Graphs page.

Here are the steps that I followed to set up the split page:

1. **Insert a Notes page. Press** ctrl I ⇨**Add Notes.**

2. **Split the screen. Press** doc▾⇨**Page Layout⇨Select Layout and choose the side-by-side option.**

3. **Move your cursor to the** Press Menu **message and press** 🖰 **to make that side of the screen active; then press** menu⇨**Add Graphs.**

Math boxes can be used to define variables and functions (and anything else that you can do on a calculator page). Because I want to use the slope formula, I am going to define the variables *x1, y1, x2,* and *y2.* Use separate math boxes (ctrl M) and type **x1:=3**, **y1:=4**, **x2:=-6**, and **y2:=-2** to define the variables.

On the Graphs page, place a point on the graph. Right-click, ctrl menu ⇨ Coordinates and Equations. Move your cursor over the *x*-coordinate and right-click, ctrl menu ⇨Variables⇨Link To⇨*x1*. See the first screen in Figure 22-5. Link the *y*-coordinate to *y1,* and repeat the process to place another point linked to *x2* and *y2.* Use the Line tool to construct a line that connects the two points.

Insert another math box and define *m* using the slope formula. See the second screen in Figure 22-5. Do you notice how the slope displays nicely as a simplified fraction? Try clicking in a math box and changing the value of *x1*. The slope automatically updates! See the third screen in Figure 22-5. The Graphs page also updates when the variables are changed.

Notes with interactive math boxes provide a dynamic way to display calculations. Math boxes can be used to interact with many other applications as well. Explore the possibilities.

Figure 22-5:
Slope investigation using Notes with interactive math boxes.

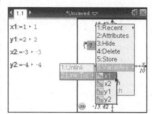 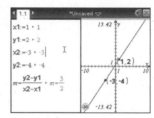 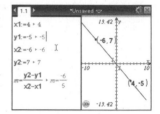

A program can be activated using a math box on a Notes page. Because of the dynamic nature of Notes with interactive math boxes, the program continually runs while the document is open. Advanced authors on the TI-Nspire use this feature to create some really intense documents.

Using chem boxes

Chem boxes are not functional; they don't perform any calculations. However, if you wish to quickly and easily display chemical formulas or chemical equations in the correct form, chem boxes are for you.

On a Notes page, press menu⇨Insert⇨Chem Box to insert a chem box. Or, press ctrl E to save a few keystrokes. To enter the symbols for calcium carbonate, simply enter **CACO3**. The chem box automatically formats the chemical compound name correctly as $CaCO_3$. Use the ⌃ key to insert a superscript. Use the ⊞ key to separate two products or reactants. Using the ⊟ in a chem box automatically inserts the yields symbol.

Part VIII
TI-Nspire Computer Software

By Rich Tennant

"What exactly are we saying here?"

In this part . . .

This part gets into how TI-Nspire's Computer software is used to support and enhance your TI-Nspire handheld experience. I show you how to use the Computer software to manage and transfer files as well as take screen shots of your device, back up files, and upgrade your handheld's operating system.

I also discuss how you can use TI-Nspire Computer software to quickly create and edit documents. I highlight two features that can only be accomplished using TI-Nspire Computer software: inserting a variety of questions into a document and adding a background image (picture) to various applications.

Chapter 23

Getting Started with TI-Nspire Computer Software

*T*I-Nspire Computer Software (and TI-Nspire CAS Computer Software) is a computer application that enables you to experience the same functionality of your TI-Nspire device in a computer environment.

For the remainder of this chapter, I refer to the numeric TI-Nspire Computer Software, not TI-Nspire CAS Computer Software. The only difference that exists between these applications is that the TI-Nspire CAS Software has a few extra commands.

Three Great Reasons for Using TI-Nspire Computer Software

Whether you are a student or an educator using TI-Nspire, you have several reasons to opt for acquiring (and using) TI-Nspire Computer Software. Here are my top three reasons:

- ✔ **Time savings:** With TI-Nspire Computer Software, you can create a document on your computer and transfer it to your handheld device. Working on your computer gives you the advantage of using a full QWERTY keyboard. Additionally, your computer mouse replaces the Touchpad on the handheld device. As much as I like the Touchpad, it simply does not compare with the mouse from a speed standpoint.

- ✔ **Easy access to computers:** Chances are that you use a computer for a few hours a day (at least). So, whether you are a student or a teacher, you may find it easier to work on a larger screen than a handheld offers. Many students use computers when they do their homework anyway, so it is a natural fit. As you see later in this chapter, files created on your computer have exact compatibility with those created on your handheld.

- ✔ **Classroom demonstration:** This reason applies more to educators rather than to individual users. More and more classrooms are equipped with computers and projection systems. If you have this technology available to you, TI-Nspire Computer Software can be projected to the entire class. This is a good option if you want to demonstrate a concept to your students. But why stop at a demonstration when the students can experience the mathematics on their own handhelds? Using the Computer Software, USB hubs (like the kind provided with Connect-to-Class Teacher Software), and cables (USB to mini-USB), teachers can transfer their .tns documents to all the students in the class. In my classroom, I use TI-Nspire Navigator to transfer .tns documents wirelessly to the students.

TI-Nspire Student Software versus TI-Nspire Teacher Software

You find two types of TI-Nspire Computer Software: TI-Nspire Student Software and TI-Nspire Teacher Software. The TI-Nspire Student Software now comes free with the purchase of a TI-Nspire Handheld. Of course, both can be purchased at instructional dealers or the TI online store.

The two types of software are extremely similar. When it comes to creating documents (called working in the Documents workspace), only one difference exists. TI-Nspire Teacher Software has an additional application called the Question application. Using the Question application, teachers can create different types of questions (custom choice, open response, and so on) and embed the question in a .tns document. Using the Documents workspace of TI-Nspire Computer Software to create documents is the topic of Chapter 24.

The TI-Nspire Teacher Software was designed to be used in a classroom setting. Consequently, it does have a few more features available. For example, TI-Nspire Teacher Software can be used to send .tns documents to multiple users at one time and to save a .tns document in a lesson bundle (with a .pdf and/or a .doc document), and it has a Content workspace to preview and download files (the topic of Chapter 25).

Installing TI-Nspire Computer Software

Before purchasing TI-Nspire Computer Software, I recommend that you download a free 30-day trial of the TI-Nspire Student Software from TI's Web site. Just go to `http://education.ti.com`, click the Downloads & Activities drop-down menu, and choose Apps, Software & Updates. Two more drop-down menus appear. For Technology, choose TI-Nspire, and for View, choose Math & Science Computer Software (click the Find button to submit).

To install a purchased version of TI-Nspire Computer Software from a CD, insert the CD and follow the prompts. I recommend that you select the default options that pop up during the installation process. A shortcut is automatically placed on your computer desktop, unless you deselect the check box for this during the installation process. I should also mention that it is important to be connected to the Internet during the installation process.

Launching TI-Nspire Computer Software and Navigating the Welcome Screen

Here are two ways to launch TI-Nspire Computer Software:

- ✓ **Create a new document:** Double-click the TI-Nspire Computer Software icon to launch TI-Nspire Computer Software.

- ✓ **Open an existing document:** To open an existing TI-Nspire document, locate the file on your computer and double-click it.

  The Welcome screen is captured in Figure 23-1. This screen allows you to quickly access all applications and recently opened documents. My favorite feature about the Welcome screen is the large preview window in the lower-left part of the screen. If you hover over the name of a recent document, the preview window shows you the first page of that document. Or, if you hover over an application icon, the preview window shows you an example of that application as well as a short description of what the application does.

  If you get to a point that you don't feel like you need to see the Welcome screen every time you open your TI-Nspire Computer Software, a fix is available. Just deselect the Always Show This at Startup check box in the lower-left corner.

The Welcome screen on TI-Nspire Teacher Software includes three additional options. You can see the View Content, Manage Handhelds, and Transfer Documents options directly from the Welcome screen.

In Figure 23-2, I show you a typical TI-Nspire Computer application screen with annotations describing its various components.

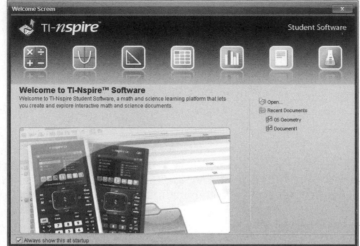

Figure 23-1:
The
Welcome
screen.

Page Sorter view

Document Toolbox | Menu bar

Tool bar

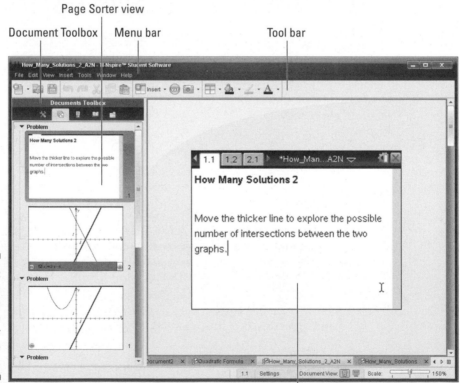

Figure 23-2:
The anat-
omy of the
TI-Nspire
Computer
Software
application.

Application work area

The .tns extension convention

All TI-Nspire files are identified by a .tns file extension. This extension is only viewable if the file resides on your computer and cannot be seen in the My Documents view on your handheld device.

While browsing for files on your computer, consider using the Detail view. By clicking the Type heading, all files are sorted by type, allowing you to quickly spot those that have a .tns extension.

Where to get your .tns files

Here are four ways that a .tns file may end up on your computer:

- ✔ You transfer a file from a handheld device to your computer using TI-Nspire Computer Software.
- ✔ You create a file using TI-Nspire Computer Software and save it to your computer.
- ✔ You mine the Internet (especially http://education.ti.com) for .tns files created by someone else and download them to your computer. Why do the work if someone else has already done it?
- ✔ A friend or colleague e-mails a .tns file to you as an attachment.

Other TI-Nspire Products

A few other TI-Nspire products are worth mentioning here. The functionality of TI-Nspire Computer Software has been expanded in recent years to include the functionality of some of these other Texas Instruments products:

- ✔ **TI-Nspire Computer Link Software:** Now obsolete, this software enables your handheld to communicate with your computer and take screen shots directly from the handheld. TI-Nspire Computer Software can do all that and more!
- ✔ **Connect-to-Class Teacher Software:** This software is no longer needed because TI-Nspire Computer Software is able to perform the function of sending .tns documents to a class set of handhelds. However, the Belkin USB hubs that usually come with Connect-to-Class are still needed to send files to more than one handheld at a time.

- ✔ **TI-Nspire Docking Station:** This is a new product that recharges class sets of TI-Nspire handhelds, transfers documents to or from multiple handhelds, and updates the OS on multiple handhelds at the same time.

- ✔ **TI-Nspire Lab Cradle:** This product can help students collect data from nearly all of Vernier's sensors. This system charges the cradles and supports users to get fast-rate data collection, including digital and multi-probe support.

- ✔ **TI-Nspire Navigator:** This is my favorite product to use with TI-Nspire Teacher Software! It allows my computer to communicate with a class-room set of TI-Nspire Handhelds wirelessly. I can transfer files, collect and grade questions instantly, take screen shots of the whole class, project a student's handheld screen . . . the possibilities are endless. It makes for a much more dynamic learning environment.

Transitioning to the Computer

The transition from working on your handheld to working on your computer is quite seamless. In the following sections, I point out some of the key differences. However, I also encourage you to open a new document and start playing around with it. You'll likely find these key differences for yourself in short order. And you'll see that these differences allow you to complete tasks with the Computer Software even more efficiently than on a handheld device.

Using your mouse

Your computer mouse replaces the Touchpad on your handheld. Here are some key tasks that you can perform with a mouse and a brief description of how these tasks differ when performed on the handheld:

- ✔ **Click and drag:** This is probably the most obvious difference. On the computer, move your mouse to an object, press and hold the left button on your mouse to select the object, and move the mouse to manipulate it. On your handheld, you must use the Touchpad to move the cursor to an object, press [ctrl][⌖] to grab it, and use the Touchpad again to manipulate the object.

  To release an object on the computer, just release the left button of the mouse. To release an object on your handheld, press the [esc] key.

- ✔ **Right-click:** As you know, I'm a huge fan of the right-click context menu. To access the context menu on the computer, move to an object or area and click the right button on your mouse. On your handheld device, use the Touchpad to move to an object or area and press [ctrl][menu].

- **Access menu items:** If you are accessing the Documents menu, just move your mouse to the menu bar and click one of the choices to view drop-down menu items. On the handheld device, you must press [doc▾] to access the Documents menu and then use the Touchpad and the [click] or [enter] key to activate the command.
- **Perform a custom split:** If you have two or more applications on a screen, move the cursor to the border of an application until the ✦ or ✢ symbol appears. Then click and drag to change the split. On your handheld device, choose [doc▾]⇨Page Layout⇨Custom Split to activate this tool. Then use the Touchpad to adjust the layout.

The Windows file management system

You can manage files located on your computer using the Windows file management system. That is, you can rename, copy, paste, delete, and so on, just as you do with other files that reside on your computer. Most of these options are available using the right-click feature. Move your cursor over a `.tns` file, click the right button on your mouse, and select an option.

TIP

Are you noticing the similarities between your computer and your TI-Nspire Handheld? Not only do you find similarities in terms of how you manage files, but much of what you have discovered about your TI-Nspire handheld also applies to TI-Nspire Computer Software as well. This is no coincidence, considering that both environments produce compatible files.

Chapter 24

File Creation and Display in Documents Workspace

* *

In This Chapter

▶ Using the Documents Toolbox to access a variety of tools

▶ Transferring the operating system to a connected handheld

▶ Deciding which view best suits your needs

▶ Locating hard-to-find tools to suit your needs

▶ Creating your own questions using the Question application

* *

*T*he Documents workspace is the next thing you see after the Welcome screen. Gaining an understanding of the menus and tools in the Documents workspace is critical if you are going to use TI-Nspire Computer Software to create your own documents.

Functionality of the Documents Toolbox

The Documents Toolbox has five different parts that perform functions vital to the creation of any document. Figure 24-1 outlines the Documents Toolbox panel. Use the tabs near the top of the panel to choose the part of the Documents Toolbox that you would like to access.

Accessing the Application menu

The Application menu should look extremely familiar. It looks and works exactly the same as the menu that you see on the handheld when you press menu. Just like your handheld, the Application menu looks different depending on which application you are working on in the Documents workspace. In the first screen in Figure 24-1, I am working on a Graphs page, so the Application menu reflects all the tools that are available in that environment.

 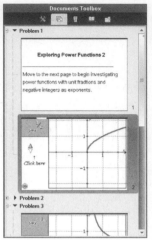 

Figure 24-1:
Location
of the
Documents
Toolbox.

Checking out the Page Sorter view

The second tab of the Documents Toolbox is the Page Sorter view. On your handheld device, press ⌃▲ for a Page Sorter view of a document. With TI-Nspire Computer Software, the Page Sorter view is organized vertically in the Documents Toolbox panel. Referring to the second screen in Figure 24-1, the Page Sorter view reveals thumbnail pictures of the two pages that comprise Problem 1. Notice that this document contains at least three problems. To reveal the contents of Problem 2 in the Page Sorter view, just double-click Problem 2. Double-click a problem number a second time to hide its contents.

When you are in the Page Sorter view, you can click any page in the slider sorter to bring the page into full view. To perform work in the full page area, you must move the mouse to the Full Page view and click once.

To change the page order, click and drag a page in the Page Sorter view and drop it to a new location. To delete a page, just click the thumbnail and press Delete.

Use the thumbnail screens to quickly navigate from one page to any other page in the document. If you would like to copy a page from one document and paste it somewhere else, both the copying and pasting are best done using the Page Sorter view. Follow these steps to copy a page:

1. **Click the thumbnail page that you would like to copy.**

2. **Press Ctrl+C to copy the page.**

3. **Click in the Page Sorter view where you would like to paste the page, and press Ctrl+V to paste the page.**

TIP

Do you want to copy more than one page of a document? Consider copying and pasting an entire problem. The steps are similar, but in Step 1, click the name/number of the problem that you would like to copy (located directly above the first page of any problem in the Page Sorter view).

Using the virtual keypad

The third tab in the Documents Toolbox is the TI-SmartView™ Emulator (or what I refer to as the virtual keypad). You have a number of different ways to display the keypad (nine in TI-Nspire Student Software). Using the drop-down menu near the top of the keypad, you can choose which calculator to display: TI-Nspire CX, TI-Nspire with Touchpad, or TI-Nspire with Clickpad. In addition, each keypad can be displayed in these three views: Normal, High Contrast, and Outline. See Figure 24-2.

Figure 24-2: Three keypad views for TI-Nspire CX.

Normal keypad   High Contrast keypad   Outline keypad

Why would you want to use the keypad when all the tools are easily accessible? You have at least two reasons. If you learned TI-Nspire on the handheld, sometimes it is easier to find the command/symbol you are looking for by typing it on the keypad. Because I know it so well, I often choose to use it to access items that are a bit tricky to find (or if I don't have the patience to try to find them). Consider, for example, the Store Variable operator ($\rightarrow$). To access this symbol on the virtual keypad, just click [ctrl] [var]. To find this symbol without using the virtual keypad, choose Utilities⇨Symbols to open the Symbols palette. Scroll down to find the Store Variable operator and press Enter to paste it into your application.

Teachers may choose to use the keypad for another reason. When I use TI-Nspire Teacher Software in my class, I usually have the keypad showing in the Documents Toolbox as a good visual aid, in case I want to point out the physical location of certain keys to my students.

Using the utilities

The fourth tab in the Documents Toolbox houses the following utilities: Math Templates, Symbols, Catalog, Math Operators, and Libraries. If you need a template, symbol, or access to a command in the Catalog, choose this tab in the Documents Toolbox. Each of these five utilities is represented in the Catalog on the handheld.

Math templates are the same ones that are found by pressing [⊞] on the handheld. See the first screen in Figure 24-3. The Symbol Palette is identical to the symbols you will find by pressing [ctrl][⊞] on the handheld. See the second screen in Figure 24-3. One thing I really like about accessing the symbol palette in TI-Nspire Computer Software is the ability to see 64 different symbols at one time (compared to 30 on the handheld). This makes it much easier to find the symbol that I want. The Catalog also has a key on the handheld that is identical, [⊞], as long as the first tab in the Catalog is chosen. See the third screen in Figure 24-3.

Figure 24-3:
Using the
Utilities
part of the
Documents
Toolbox.

Math Templates

Symbols

Catalog

The Math Operators utility organizes commands by topic. I occasionally use this if I forget the name of a command that is associated with a particular topic (such as Matrices commands). The last utility, Libraries, is connected to the Program Editor. See Appendix C for more details about this utility.

Transferring files (and operating systems) with Content Explorer

This tool is easy to use! If you want to get a file from your computer to your handheld (or vice versa), you can use the last tab in the Documents Toolbox, Content Explorer. Connect your handheld to your computer using a USB–to–mini-USB cable. Almost immediately, the Handheld File browser, located in the lower half of the Content Explorer panel, will recognize your handheld with an icon. (See Figure 24-4; my icon is labeled TI-Nspire 3C9E.) Double-click the icon and locate the folder that you would like the file to be transferred to. Now, locate the file (using the Computer File browser in the upper half of the Content Explorer panel) on your computer that you would like to transfer. Use your mouse to drag and drop the file. If you would like to transfer an entire folder, just drag and drop the folder. In this way, you can easily transfer files back and forth between your computer and your handheld. See an expanded view in the second screen in Figure 24-4.

 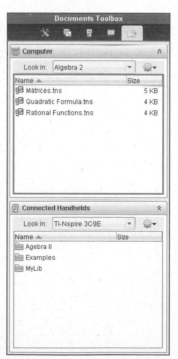

Figure 24-4: Using the Content Explorer panel of the Documents Toolbox.

This tool can be used to transfer a new operating system as well. I like that when you double-click on a connected handheld, the software will prompt you to update the OS if the OS on the connected handheld is not as up to date as the OS on your computer software.

Much like the toolbar on your computer, the Documents Toolbox can be hidden from view. To accomplish this, choose Window➪Autohide Documents Toolbox. Just click on the words *Document Toolbox* that appear on the dock on the left side of your document workspace to show the Documents Toolbox.

TI-Nspire Computer Software Views

When using TI-Nspire Computer Software, two different views are available. They can be accessed by clicking the View drop-down menu and choosing Handheld View or Computer View. You have a few other ways to access these views. For Handheld view, press Alt+Shift+H or click the small Handheld icon near the lower-right corner of the screen. For Computer view, press Alt+Shift+C or click the small Computer icon near the lower-right corner of the screen. See Figure 24-5.

The Handheld view

This view is identical to the screen that you see on your handheld. I often use this view when projecting to the whole class if my students are simultaneously working on the document on their handhelds. I want them to see the same thing at the front of the classroom that they see on their handhelds. One thing that I like to do when I am in Handheld view is to adjust the scaling to make the screen larger. In the status bar (near the lower-right corner of the screen) is a slider bar that's set to the default of 150%. Simply grab and move the slider to adjust the screen size accordingly. In Figure 24-5, the Scale slider is set to 200%.

The Computer view

When a screen gets too crowded and you can't see as much of the screen as you would like, it is a good time to use the Computer view. As can be seen in Figure 24-6, the Computer view offers quite a bit of space with which to work. In fact, this view is the likely choice if you want to include four applications on a single page.

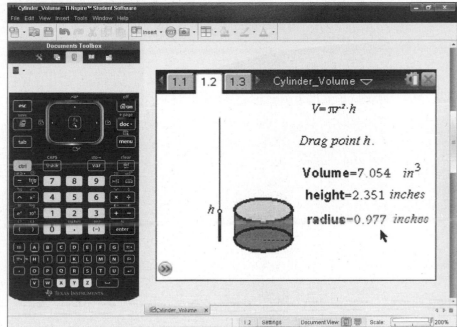

Figure 24-5:
The
Handheld
view scaled
to 200%.

Did you notice that the Scale slider has been replaced with a Boldness slider? The drawback of working in Computer view is that it can be difficult to read the thin font and see all the thin lines on a Graph page. Of course, a fix is available. Try adjusting the Boldness slider to change the thickness of the fonts/lines.

Here is a side note worth mentioning. Originally, the Boldness feature was not available on TI-Nspire Computer Software. However, Texas Instruments conducts focus groups with teachers. TI asks for ideas to improve its products, and it listens to the responses. The company often uses these responses to improve its products. The Boldness feature is just one of many such changes.

When you are in Computer view, the ten navigation keys on the upper part of the virtual keypad will not work.

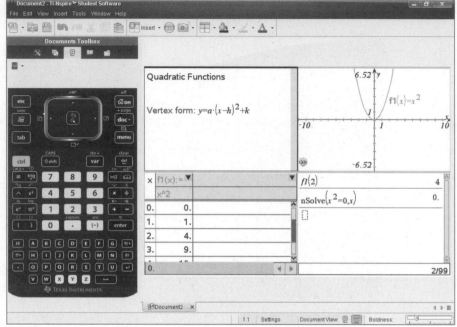

Figure 24-6:
Using
Computer
view to
display up to
four appli-
cations on
one page.

Using the Workspace Toolbar

Tools are easily accessible from the Workspace toolbar (right below the
menu bar at the top of the screen). See Figure 24-7.

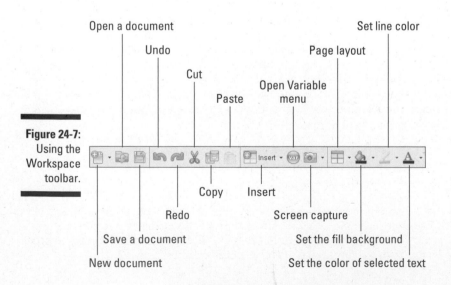

Figure 24-7:
Using the
Workspace
toolbar.

Creating, opening, and saving a document

Click the first icon on the Workspace toolbar to create a new document and select New TI-Nspire Document (press Ctrl+N). I address the second choice, Create PublishView Document, later in this chapter.

If you want to open an existing document, click the folder icon and use the drop-down menus to locate and select the file that you would like to open.

If you would like to limit your search to .tns files, click the last field, Files of Type, and choose TI-Nspire Documents (*.tns). See Figure 24-8.

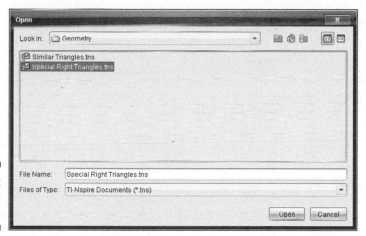

Figure 24-8:
Opening a
file.

If you open multiple TI-Nspire documents, the document titles will appear in tabs near the bottom of the screen. To change documents, click a different tab. To close a document, click the X on the tab you would like to close.

To save a document that has not been previously saved, click the Save icon located on the toolbar or choose File⇨Save Document. Type a filename and determine the location to which you want to save the document.

To save your document to a new location or with a different name, choose File⇨Save As and follow the same steps I just described. This is especially helpful if you are trying something that you are not sure will work. You will still have the original file to fall back on.

After saving a document, periodically press Ctrl+S to save your work to the same filename and location. Keep in mind that this process overwrites the last saved version of the current document. Saving periodically helps protect you from the devastating effect of computer crashes.

Shortcuts for editing documents

The next five shortcut icons allow easy editing of your .tns files. If you have some experience working on a computer, you are probably quite familiar with how to use these editing commands: Undo, Redo, Cut, Copy, and Paste. I like the fact that I don't have to use a drop-down menu to find any of these commands. Just click the familiar icons and edit away!

Understanding the Insert drop-down menu

Here is a brief description of each of the items that can be inserted:

- **Insert page:** Choose Insert⇨Page (or use the shortcut key sequence Ctrl+I) to insert a new page after the current active page. Then, select the application for the page that you want to insert.

- **Insert problem:** Choose Insert⇨Problem to insert a new problem after the current active problem. You are prompted to select an application for the first page of the new problem.

- **Insert (choose your application):** If you know the type of application that you want to add to your new page, click Insert and select the desired application for the page.

- **Insert image:** Choose Insert⇨Image to insert an image in the background of a Graphs page. You must choose the location of the image to be displayed (only .jpg, .jpeg, .bmp, and .png files can be selected). See examples of this feature in the color section of this book.

- **Insert text box:** Choose Insert⇨Text Box to insert a text box when working on a PublishView document.

- **Insert Program Editor:** Choose Insert⇨Program Editor to insert a Program Editor page. If you are on a Calculator page, it will automatically split the screen; otherwise, it will give you a full-page view of the Program Editor. See Appendix B for more details.

- **Insert Sensor Console:** Choose Insert⇨Sensor Console (or use the shortcut key sequence Ctrl+D) to insert the Sensor Console bar at the bottom of the screen. This is used in conjunction with data collection (see Chapter 20).

Calling up variables, taking screen shots, and changing the layout

Press the circular Variable icon to quickly access any of the variables or functions that have been stored in the problem you are working in.

Press the Camera icon to take screen shots. Screen captures are also available from the Tools drop-down menu. You have two options:

- ✔ **Standard screen capture:** This takes a screen shot of the application area screen on your computer. A small notification appears in the lower-right portion of the screen that says `Screen Capture Taken. View It`. Click View It to open the Page Capture window (also available on the Window drop down menu). There, you can edit and save your screen capture. (I like the fact that you can save your screen capture to many different file types.) See Figure 24-9.

- ✔ **Capture selected handheld:** If your handheld is connected to your computer, the Screen Capture tool takes a screen shot of the screen on your handheld.

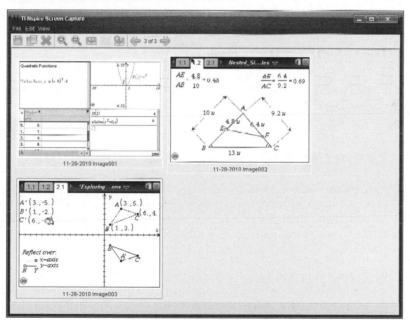

Figure 24-9:
The Page
Capture
window.

 Click the screen that is divided into four parts to change the page layout. Use this tool to split the screen into up to four different parts.

Adjusting the color

The final three icons on the Workspace toolbar are for adjusting the color of text, objects, and lines. These are easy tools to use, just highlight the text, object, or line and then click on the tool icon to change the color.

The icon that looks like a paint can will highlight text with a background color. This can be used in Notes (including math boxes) and Lists & Spreadsheet. In the Graphs, Geometry, and Data & Statistics environments, this tool will fill a shape with color.

 The icon that looks like a paint brush will change the color of a line, point, or object in a Graphs or Geometry environment.

 The icon that looks like the letter A will change the color of selected text in a Notes (including math boxes) or Lists & Spreadsheet environment.

 A sneaky way to change the color of text on a Graphs or Geometry page is to place a point on the page. Then, right-click and change the points attributes to Thin and change its color to whatever you want. Now, right-click and label the point; any text you type will match the color of the point!

Hidden Tools on the Menu Bar

Some tools are harder to find than others. It would be easy to overlook some of these tools, but I think you will agree that some of them are quite useful.

Creating a PublishView document

PublishView is not really a view, it is a document type. Using PublishView, you can place multiple dynamically linked TI-Nspire pages on the same sheet along with objects like text boxes, hyperlinks, and even video! To open a new PublishView document, choose File⇨New PublishView Document (or press Ctrl+Shift+N). Saving a document creates a .tnsp file.

Existing .tns files can be converted to PublishView documents by choosing File⇨Convert⇨PublishView Document. On the flip side, files created as PublishView documents can be converted to .tns files by choosing File⇨Convert⇨TI-Nspire Document.

PublishView is not line-based, like a word document . . . it is frame based. You can add frames (and even overlap frames) by dragging and dropping one of the document tool "boxes." There are eight TI-Nspire applications you can drag and drop and four different PublishView objects. The PublishView options include an image, a video, a text box (customize the font), and a hyperlink. Each frame's size can be customized, just click on it and drag the corners.

You don't need to have TI-Nspire software to view a PublishView document. Both PublishView and TI-Nspire documents can be exported to Web pages or HTML snippets that can be used to paste into existing Web pages or blogs. This feature is powered by TI-Nspire Document Player (which works in conjunction with but is different from TI-Nspire Computer Software). Additionally, TI-Nspire documents can be exported to a file that is compatible with Cabri. Hidden in the File drop-down menu is the Export option; choose File⇨Export.

You can embed a PublishView document in any environment that supports HTML, including Web pages, PowerPoint, Moodle, Blackboard, and so on. The ability to include TI-Nspire documents in different environments makes this a really exciting new feature. I like showing multiple representations in my class; PublishView takes that to a whole new level! See Figure 24-10 to see an example of part of a lesson in PublishView

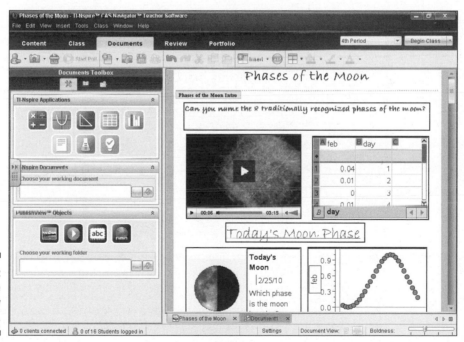

Figure 24-10:
The
PublishView
window.

Printing, saving, and changing the settings

You can print the contents of a TI-Nspire document. Choose File⇨Print and customize to fit your needs. I highlight three items that I like to customize when I print a TI-Nspire document (see Figure 24-11):

✔ Select an option from the Print What field:

   • Viewable Screen: Prints the viewable window with no scrolling.

   • Print All: Prints any text or data that you have to scroll to see.

✔ Customize the Pages per Sheet field: 1, 4, and 8 are available options.

✔ Deselect the Group Pages by Problem check box to save paper.

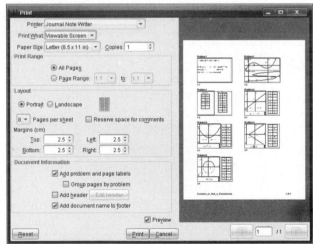

Figure 24-11:
Printing options.

So far, I have talked about the Save command and the Save As command. On the File menu is a command called Save to Handheld; this accomplishes exactly what it says. If a handheld is connected to your computer, the Save to Handheld command is no longer grayed out, and you can use it to save your TI-Nspire document directly to your handheld.

You can save a file as read only if you don't want anyone to change the file you saved. To accomplish this, click File⇨Document Properties⇨Protection and check the box that says, "Save this document as Read Only."

In TI-Nspire Computer Software, you have two additional locations from which to access the settings. One way to access the settings is to click the File drop-down menu and choose Settings. Or, you can move your mouse near the bottom of the screen, look for the word *Settings* on the status bar, double-click the word, and customize your settings to your liking.

Question, the Other Application

The Question application is important if you are teaching in a classroom. The Question application is not an available choice if you add a page — it can be accessed only from the Insert drop-down menu. Using the Question application, questions can be embedded in your TI-Nspire document. The Question application is available only in TI-Nspire Teacher Software.

Even if you are just starting to find out how to use TI-Nspire, inserting questions into documents is an easy thing to do. You may be surprised to see all the different question types that are available. Choosing Insert⇨Question opens the Choose Question Type dialog box. Each time you use your mouse to click an option, a brief description and a small template appear near the bottom of the dialog box.

You find four main categories: Multiple Choice (7 options), Open Response (2 options), Equations (2 options), and Coordinate Points & Lists (3 options). That is a total of 14 different types of questions that can be inserted into a TI-Nspire document.

If you look closely at Figure 24-12, you will notice four different pages of questions that are listed on the same screen in TI-Nspire Teacher Software. To accomplish this, I chose Window⇨Show Document in Tabs. Each question came from a different document and is shown in a thumbnail view.

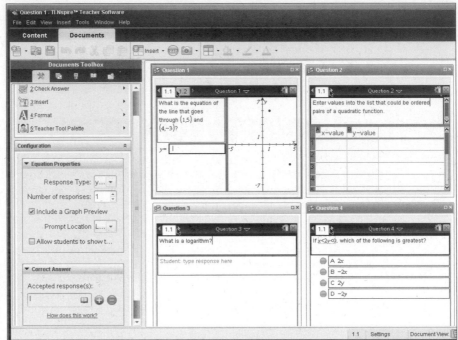

Figure 24-12:
Inserting
questions.

Chapter 25

File Management with Content Workspace

*T*he Content workspace is available only in TI-Nspire Teacher Software. With the Content workspace, teachers can manage content and easily search (and preview) new activities online. It is also designed to increase connectivity with student handhelds by using USB cables or the TI-Nspire Docking Station. I am all for any tool that makes a teacher's job easier!

To access the Content workspace, click the tab near the upper-left corner of the screen. See Figure 25-1 for the basic layout of the Content workspace.

On the left side of the screen is a rectangular panel labeled Resources. This area is where I do most of my work in the Content workspace. I use it to quickly locate and preview files.

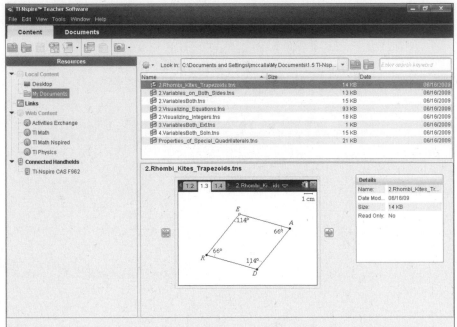

Figure 25-1:
The Content
workspace.

Accessing Local Content

Any content that is saved on your computer can be accessed by using the Local Content shortcuts in the Resources panel. By default, you are given two choices, Desktop or My Documents. If you click either choice, the associated folders and files display in the panel in the upper-right part of the screen. I recommend choosing My Documents⇨TI-Nspire. Of course, to make this selection, you must temporarily move your cursor to the upper-right panel and click the TI-Nspire folder to select it. In this folder, you should find some folders that contain .tns files.

Wait! Didn't I say that using the Content workspace saves time? Here's how. You can add your own link to frequently used folders or files. Follow these steps to add a link of your own to the Local Content section:

1. **Right-click the folder that you would like to add a link for (or right-click the My Documents option below Local Content).**

2. **Choose Create Shortcut.**

   Notice that the selected folder is now located in the Local Content part of the Resources panel. Whenever I want to access a .tns file in my Algebra II Honors class, I save time by accessing the document through

Local Content links. See Figure 25-2. Using this technique, I avoid searching through folders and submenus to find the file I need and to maximize the time that I have with students.

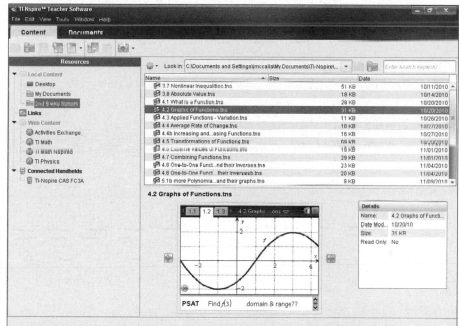

My favorite feature of the Content workspace is the preview panel in the lower-right part of the screen. Using Local Content, open a folder and click a .tns file once. In the preview panel, you can see the first page of the TI-Nspire document you selected. Using the small arrows to the left and right of the preview screen, you can view every page of the file without ever having to go to the trouble (and time) of opening the file! I have hundreds of TI-Nspire files on my computer. At times, I have opened over 20 documents at one time to get to the one that I need. From experience, it takes a long time to close 20 documents, but never again! Using the Preview feature saves me time and energy.

Using Links to Launch Web Sites

The next part of the Resources panel is the Links option. Links are an easy way to visit often-used Web sites. By default, all the links are to different Texas Instruments Web sites that contain TI-Nspire content. However, by clicking the Add Link icon right above the list, you can add a link of your own (or right-click the Link option and choose Add New Link).

How does this help you as a teacher? Well, you can launch a Web site without ever having to leave your TI-Nspire Teacher Software platform! No more bookmarking Web sites on your Web browser (my list of bookmarks is already way too long).

Investigating Web Content to Preview Activities

The next part of the Resources panel is the Web Content selections. Clicking these Web sites provides a different type of preview. For example, I clicked Activities Exchange and then chose an activity called *Mathematical Superlatives: An Exploration of Optimization Problems.* In the preview panel, I see a short description of the activity. All the files associated with the activity are in the small box in the lower-right corner of the screen. See Figure 25-3. I can double-click any of these files to open the file (`.doc`, `.pdf`, and `.tns` are the most common). If I like what I see, I can click the Save This Activity to Computer button on the right side of the screen. All files associated with the activity will be saved as a lesson bundle.

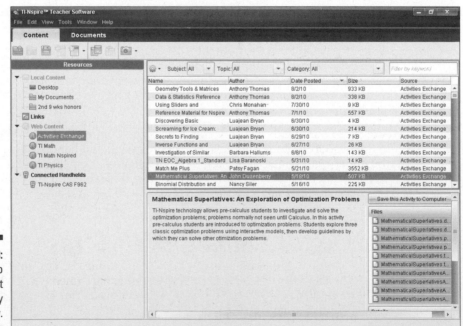

Figure 25-3:
The Web Content activity preview.

Creating Lesson Bundles

What is a lesson bundle? Think of it as a zipped file. Different types of documents (usually `.pdf`, `.doc`, and `.tns` files) are bundled together to allow easy distribution among other educators. When you see a file that ends in `.tilb`, you know that you are looking at a lesson bundle. Lesson bundles work seamlessly with TI-Nspire Teacher Software. If you send a lesson bundle to connected handhelds, only the `.tns` files will transfer.

To make a lesson bundle of your own, click the Create a Lesson Bundle icon right below the Content tab (or right-click any file and choose Lesson Bundles⇨New Lesson Bundle). Type the name of your bundle in the highlighted area and then right-click the bundle (choose Lesson Bundles⇨Add Files to Lesson Bundle).

Using Connected Handhelds to Transfer Files

The last part of the Resources panel is only available when handhelds are connected via USB cables. I have used the Belkin USB hubs that I got with my Connect-to-Class Teacher Software to connect up to seven handhelds to my computer at a time. (Actually, you could successfully connect more handhelds at once, but I think it gets a little crowded when more than seven students are standing around my computer.) If you have a classroom set of handhelds, you might consider using the TI-Nspire Docking Station to connect up to 40 handhelds at one time (and avoid overcrowding issues altogether).

The transfer tool or the TI-Nspire Docking Station can be used to send a new OS to many handhelds all at once. That is a big time-saver!

After your handhelds are connected, click the Connected Handhelds option in the Resources panel, and a list of all the connected handhelds appears in the upper-right panel. This list is very informative. You can see the percentage of battery remaining (both TI-Nspire Rechargeable and AAA batteries) for each of your handhelds. In addition, the OS number and available storage are provided. This makes it easy to see whether a student needs to upgrade his or her OS.

Each handheld in the list is given a name. If you right-click the name of a handheld, one of the options is to rename the handhelds. You could rename the handhelds using student names so that you know when a student is connected. See Figure 25-4 to see other available information.

Figure 25-4:
Right-click
to rename
your
connected
handhelds.

After the handhelds are connected, you can quickly and easily transfer files from your computer to the connected handhelds. Follow these steps to transfer files:

1. **Use the Resources panel to locate the file(s) or folder(s) that you would like to transfer.**

2. **Click a file/folder to highlight the file that you would like to transfer.**

   If you would like to transfer more than one file/folder, hold down Shift (to select multiple consecutive entries) or hold down Ctrl (to make multiple selective entries) to make your selections.

3. **Click the icon shown here and choose Send to Selected Handhelds to open the Transfer tool window.**

   You can add even more files to the transfer list by clicking the Add to Transfer List button.

4. **To activate the transfer, click the Start Transfer button.**

   After the transfer has been activated, the Transfer tool continues to send out files to any handheld that connects to it. Only after you press the Stop Transfer button will your computer stop sending out the files. See Figure 25-5. Another way to activate the Transfer tool is to right-click a file and choose Send to Handhelds. Alternatively, press Ctrl+Shift+T to activate the Transfer tool.

When I use this feature in the classroom, I often leave the Transfer tool running in the background during the whole class period. That way, students who come in late can plug the mini-USB cable into their handheld and have the documents seconds later!

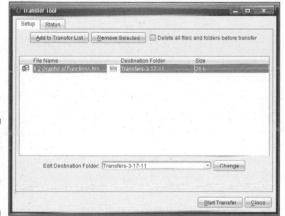

Figure 25-5:
Transferring files using the Transfer tool.

Sorting and Backing Up Your Files

If you ever have trouble finding a document in a long list of TI-Nspire files, a few tools are available that come in really handy. A Search box is located in the upper-right corner of the screen. Type a keyword to use the Search tool. I think that the Sort feature can also be helpful in locating tough-to-find files. Click the View drop-down menu and select Sort By. You find three sorting options: Name, Size, and Date.

One final note when working with TI-Nspire Computer Software (or computers in general for that matter): Make backup copies of the TI-Nspire documents that you create. I use a flash drive and save the My Documents folder periodically by choosing My Documents⇨TI-Nspire. If you have ever had your computer crash (I have!), you know how important it is to have backups of your important files.

Part IX
The Part of Tens

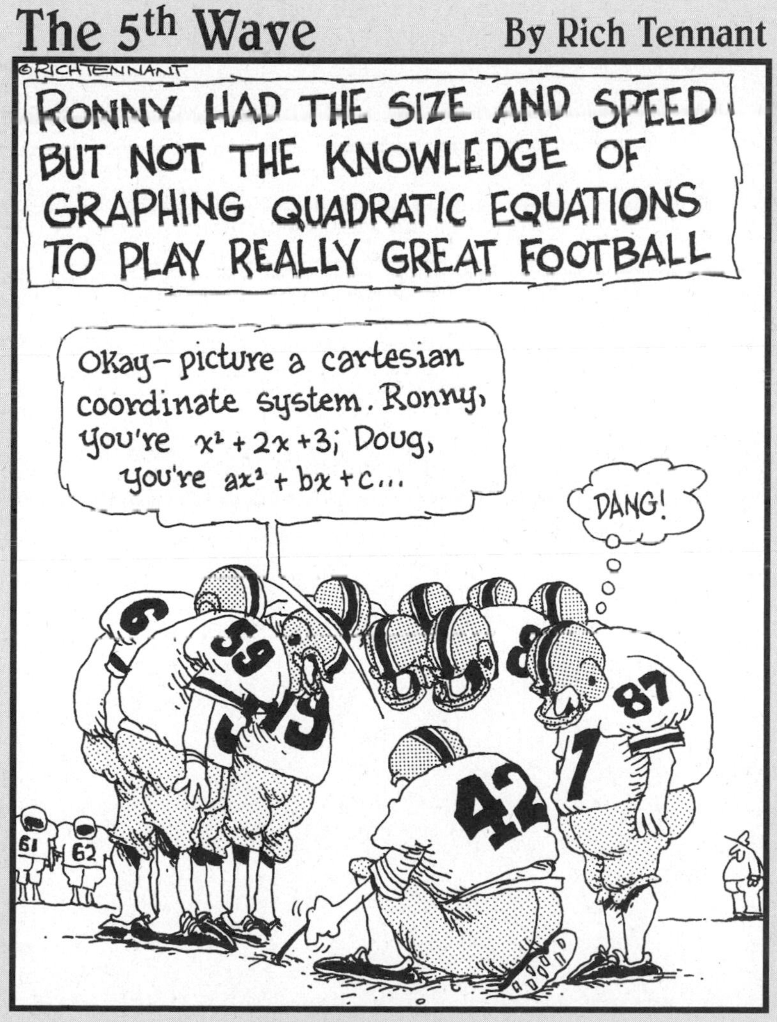

In this part . . .

This part is one of my favorite sections of the book. Here I get to pull together a bunch of great ideas (in packages of ten) as well as some warnings about what to watch out for. These great ideas come in the form of ten tips and shortcuts that are sure to get you thinking about taking your TI-Nspire experience to the next level.

I also share ten common mistakes and issues that I see in the classroom. It is my hope that you will avoid these same pitfalls by reading this chapter. In addition, I show you how to access the vast array of resources that are available on the Internet.

Chapter 26

Ten Great Tips and Shortcuts

* *

*N*o doubt, TI-Nspire is a powerful machine. Fortunately, Texas Instruments has put a lot of effort into developing a keypad, navigational tools, and a menu structure that allow you to harness this power and create documents quickly and efficiently.

In this chapter, I talk about some of the additional shortcuts and time-savers that may not be apparent to a new TI-Nspire user. Most of these shortcuts are consolidated from other areas of this book.

One Great Time-Saver

There's no substitute for working with a computer. Consider purchasing the TI-Nspire Student or Teacher Software. (TI-Nspire Student Software comes free with the purchase of a TI-Nspire Handheld.) Create your documents using this computer software and transfer them to your handheld device using a standard A–to–Mini-B USB cable. You will save an incredible amount of time, especially if you are just learning TI-Nspire or working with complex constructions. My students like having the ability to do homework on a larger screen.

Adding and Subtracting to the Decimal Place of an Ordered Pair

The level of precision of an ordered pair is determined by the settings of your TI-Nspire. If you want to change the level of precision for a specific ordered pair, hover your cursor over the *x*-coordinate so that the ⟳ symbol appears and the coordinate blinks. Press the ⊞ repeatedly to increase the number of displayed digits or press the ⊟ key to decrease the number of displayed digits. Repeat this process for the *y*-coordinate as well as for any other measurements located on Graphs or Geometry pages.

Don't forget that the coordinates themselves are clickable regions. As you hover the cursor over the *x*-coordinate, try clicking (🖰) twice to allow editing of the *x*-coordinate. Type a new *x*-coordinate value and press [enter]. Watch the point jump to its new location. You can also edit the *y*-coordinate. For functions that are not one-to-one (meaning that in some instances the *y*-values are not unique), TI-Nspire jumps to a point with the specified *y*-value closest to the current location.

Locking Variables to Protect the Integrity of Your Document

TI-Nspire provides a dynamic learning environment where variables and functions can be explored and manipulated. Sometimes, to protect the integrity of your document, you may want to limit the changes that can be made to your document. For example, graph the function $y = (x + 1)^2 - 2$ on a Graphs page. One of the built-in features of TI-Nspire is to be able to grab and move a quadratic function that is written in vertex form. But what if you don't want anyone to be able to change your function? Here's how to protect your document to prevent changes to a variable or function:

1. **To insert a Calculator page, press** [ctrl][I]⇨**Add Calculator.**

2. **Use the Lock command to protect your function by typing** Lock(f1).

If you try to grab and move the function on a Graphs page, you will get the message `Cannot move: dependent variable is locked`. In this way, you have successfully limited the dynamic environment of TI-Nspire to suit your needs. If you change your mind and want to change your function, just type **Unlock(f1)** on a Calculator page.

Using the Save As Feature

If you are working on a document and you want to add something that you are not sure will work out, press [doc▾]⇨File⇨Save As to invoke the Save As command. Choose a different filename, and then try out your idea. If it doesn't work out, press [🏠on]⇨My Docs and open the original file, which brings you back to where you were before you tried out this new idea. The Undo feature is a similar option, but it may take longer to restore your position than using the Save As feature.

Get Advanced Authoring Techniques from the Experts

This book does not address some of the tips and tricks that advanced authors on the TI-Nspire use to make the magic happen (look at some of the Math Nspired documents at mathnspired.com If you don't know what I am referring to). To create these amazing documents, these authors rely on an assortment of techniques (like using a When statement on a Graphs page). Even if you have no interest in using advanced authoring techniques to liven up your documents, the following Web sites are great resources that I think you will enjoy:

- ✔ **Stephen Arnold's Web site (**http://compasstech.com.au**):** Click Resources and Support for the TI-Nspire platform, and then choose TI-Nspire Authoring Support. Stephen is from Australia, and he has posted some Jing tutorial videos with accompanying .tns files.

- ✔ **Nelson Sousa's Web site (**www.nelsonsousa.pt**):** Nelson is from Portugal, and in the Tips & Tricks section of his Web site, you can find PowerPoints and step-by-step instructions to guide you through the creation process.

- ✔ **Marc Garneau's blog (**http://web.me.com/piman2/PimanNspire/Blog/Blog.html**):** Marc (a.k.a. Piman) is from Canada; he has posted Tips 'n' Tricks instructions to help get you started.

- ✔ **Texas Instruments Web site (**http://education.ti.com**):** Here you can find links to other advanced authors' Web sites, like Sean Bird's, Bryson Perry's, John Hanna's, and Tom Reardon's. Choose Products⇨TI-Nspire Technology⇨TI-Nspire Communities.

How to Find and Download Great Activities

If you are motivated to find more Action Consequence documents or activities like those featured in the previous section, I encourage you to visit TI's Activities Exchange. Follow these steps to find and download an activity:

1. **Go to TI's Web site at** http://education.ti.com.

2. **In the Download & Activities drop-down list, click Activities.**

3. **Search for an activity using one of these options:**

   - *Search by Subject:* Click through a series of submenus to refine your search: Technology, Subject, Subject Area, and Topic. For example, I clicked TI-Nspire Technology and found 1,211 different activities that I could download. I recommend refining your search a little more than I did!

   - *Search by Keyword:* Activities that correlate to the keyword that you provide are displayed. I did a search for Circle activities and found 306 results. If you click the drop-down menu and choose Group by Technology, you can further refine your search.

   - *Search by Standard:* Choose your state-specific standard, subject area, and grade level. This is a nice feature if your class has an end-of-course state test associated with it.

   - *Search by Textbook:* You find over 200 different textbooks listed. Select your grade level and subject.

   - *Search by TI Website:* Click one of five choices: TI Physics, Math Nspired, TI Math, TI Middle Grades, and TI Science Nspired. These activities have all been reviewed, and you can be assured that they are of excellent quality. Clicking the drop-down menu allows you to search by Most Downloaded (last 60 days) or Most Recommended (last 60 days), among other choices.

   - *Advanced Search:* Click the Advanced Search link, fill in the information, and click Search to display the activities.

4. **Click the desired activity from the list of displayed activities.**

   You are directed to a page that gives an activity overview and a list of downloadable files as well as some other useful information.

5. **If you are searching for a TI-Nspire activity, download the associated** .tns **file. Also, download any additional files, including** .pdf **and** .doc **files.**

   If I like the activity as is, I usually download the .pdf file. However, if I want to customize the activity by adding or subtracting material, I download the .doc file.

6. **Transfer the** .tns **file to your handheld device using TI-Nspire Computer Software or use the software to open the** .tns **file.**

   Don't forget to read the documentation that accompanies the .tns file.

Save Time by Copying an Entire Problem

You may have just found an excellent activity from TI's Web site, but you only want to use some of the pages of the .tns file you downloaded. Instead of painstakingly copying those pages one at a time, you can use another method. Here are the steps to copy an entire problem from one .tns file to another:

1. **Go to the Page Sorter view and press** ctrl ▲.

2. **Click on the problem name/number.**

3. **Press** ctrl C **to copy the problem.**

4. **Go to the Page Sorter view of the document that you would like to paste the problem into and press** ctrl V.

I often use this tip to prepare lessons for my students. I love being able to pick and choose the parts of a .tns file that I like and adapting them to make my classroom a more dynamic learning environment.

Join Google Groups to Get Your Questions Answered

If you have questions, something is not working as you expected, or you want to learn from the expertise of other TI-Nspire users, join the online Google Group. Follow these steps:

1. **Go to Google's Web site at** www.google.com.

2. **In the More drop-down list, click Groups.**

3. **In the Search for a Group box, do a search for TI-Nspire.**

   You'll find over 750 members worldwide at tinspire@google groups.com.

4. **Click the TI-Nspire Group and you will have access.**

   Consider joining and being a part of the discussion. You can receive e-mails every time a discussion post is made or get a daily digest of all the discussion posts made each day (the more popular option).

This online community is one of the most helpful that I have ever been a part of. Hundreds of files have been posted to the group, but the discussion section is what most people find so beneficial. Both students and teachers pose and answer questions. You will be surprised to hear from people who are from all over the world (who post at all hours of the day).

Five Sets of Wonderful Shortcuts

In the following sections, I feature five sets of additional shortcuts that are sure to improve your efficiency. The shortcuts are grouped by task and are taken from the resources provided by TI with your TI-Nspire purchase. Within each task are several shortcuts associated with that particular task.

Shortcuts for editing text

Here are several editing shortcuts, the same ones that are used with most computer applications:

- ✔ **Cut:** ⌈ctrl⌉ ⌈X⌉
- ✔ **Copy:** ⌈ctrl⌉ ⌈C⌉
- ✔ **Paste:** ⌈ctrl⌉ ⌈V⌉
- ✔ **Undo:** ⌈ctrl⌉ ⌈Z⌉ (or ⌈ctrl⌉ ⌈esc⌉)
- ✔ **Redo:** ⌈ctrl⌉ ⌈Y⌉
- ✔ **Force approximate results:** ⌈ctrl⌉ ⌈enter⌉

Shortcuts for managing documents

Here are shortcuts to access the Documents and Context menus as well as six shortcuts to help you manage documents:

- ✔ **Access the Documents menu:** ⌈doc▾⌉
- ✔ **Access the Context menu (right-click):** ⌈ctrl⌉ ⌈menu⌉
- ✔ **Open a document:** ⌈ctrl⌉ ⌈O⌉
- ✔ **Close a document:** ⌈ctrl⌉ ⌈W⌉
- ✔ **Create a new document:** ⌈ctrl⌉ ⌈N⌉
- ✔ **Insert a new page:** ⌈ctrl⌉ ⌈I⌉
- ✔ **Select an application:** ⌈ctrl⌉ ⌈K⌉
- ✔ **Save the current document:** ⌈ctrl⌉ ⌈S⌉ or ⌈ctrl⌉ ⌈▤⌉

Shortcuts for accessing symbol palettes

Here are shortcuts for inserting tough-to-find characters and symbols in a document:

- Display the Character/Symbol palette: ⟨ctrl⟩⟨⌨⟩
- Display the Expression Template palette: ⟨▦⟩
- Display the Trig Symbol palette: ⟨trig⟩
- Display the Π Symbols palette (Π, ∞, θ, and so on): ⟨π▸⟩
- Display the Equality/Inequality palette: ⟨ctrl⟩⟨=⟩
- Display the Marks and Letter Symbols palette (?!$°:';_\): ⟨?!▸⟩

Shortcuts to modify the display

These secondary key shortcuts help you adjust the contrast and turn off the device:

- Increase contrast: ⟨ctrl⟩⟨+⟩
- Decrease contrast: ⟨ctrl⟩⟨−⟩
- Power off: ⟨ctrl⟩⟨⌂ on⟩

Shortcuts that are application specific

Most of these shortcuts do not work in every environment:

- Hide/show entry line (in Graphs or Geometry): ⟨ctrl⟩⟨G⟩

  This is also a Go To shortcut in Lists & Spreadsheet and the Program Editor.

- Insert math box (in Notes): ⟨ctrl⟩⟨M⟩
- Open the Scratchpad (in any application): ⟨📓⟩ or ⟨ctrl⟩⟨O⟩
- Add a function table (in Lists & Spreadsheet, Graphs, or Geometry): ⟨ctrl⟩⟨T⟩
- Group/ungroup (in any application): ⟨ctrl⟩⟨4⟩ / ⟨ctrl⟩⟨6⟩
- Check Syntax & Store (in the Program Editor): ⟨ctrl⟩⟨B⟩

Computer Keyboard Shortcuts

Sometimes when I am using a computer, I get into typing mode. Call it laziness if you want, but I don't really feel like accessing a symbol palette or catalog to insert the symbol or command that I need. When you press [enter] to evaluate the command, the symbol or command looks exactly as it should had you not been so lazy. Incidentally, the first seven shortcuts also work on the handheld:

- **To enter π, type** pi.
- **To enter θ, type** theta.
- **To enter ∞: type** infinity.
- **To enter ≤, type** <=.
- **To enter ≥, type** >=.
- **To enter √, type** sqrt(...).
- **To enter | |, type** abs(...).
- **To enter *i* (the imaginary constant), type** @i.
- **To enter *e* (natural log base *e*), type** @e.
- **To enter ° (degrees), type** @d.
- **To enter ʳ (radians), type** @r.

Chapter 27

Ten Common Problems Resolved

· ·

1 f you are like I am, mistakes are part of the learning process when it comes to getting to know a new tool such as TI-Nspire. Some of the mistakes that you make with TI-Nspire result in an error message. Other mistakes are a bit more insidious — it's not necessarily that you've done something wrong; it's just that the device isn't doing quite what you want it to do. I talk about both types of issues in this chapter. You can be assured that the material presented in this chapter comes strictly from personal experience in my classroom. It is my hope that by sharing the issues that my students and I have dealt with during the learning process, you may be spared some aggravation.

TI-Nspire is so similar to a computer that many of the methods and procedures you have acquired while working on a computer can be applied to TI-Nspire. You can access a variety of shortcuts to work efficiently, navigate documents using a variety of computer-like actions (such as Page Down), and save documents using methods that are virtually identical to those for a computer.

My favorite computer-like feature happens to be my number-one remedy when it comes to dealing with mistakes: the Undo feature. Just as with a computer, you can press the shortcut key sequence [ctrl][esc] or [ctrl][Z] repeatedly to back out through a series of steps. The Undo feature fixes the majority of problems, or it at least gets you back to a point where you can try a different tactic. If you back out through a series of steps only to realize that you were correct, you can use the Redo feature ([ctrl][Y]) to move forward through the steps that you've just undone.

The following sections contain a list of ten specific errors or problems and describe their accompanying solutions.

Accessing the Maintenance Menu to Resolve Problems

Occasionally, things go wrong on the TI-Nspire. If your handheld doesn't turn on, the key presses don't respond properly, or the display is garbled, you need to take action. I usually try putting new batteries in first, but if that doesn't work, follow these steps to access the Maintenance menu:

1. **Remove the keypad and remove one battery from each column of batteries for 2–5 seconds.** (Not applicable to TI-Nspire CX)

2. **With the handheld turned off, follow the steps below:**

   • **TI-Nspire CX: Press [doc▾] [EE] [enter], and then, while holding these keys, use a pen to press the Reset button on the back of the handheld.** This maneuver takes a lot of agility! I tried putting the pen in my mouth to press the Reset button. I recommend recruiting a friend to help.

   • **TI-Nspire Touchpad: Press [doc▾] [EE] [enter], and then, while holding these keys, press [⌂ on].**

   • **TI-Nspire Clickpad: Press c·P then w.**

The TI-Nspire loads the Maintenance menu. Here are the four options from which you can choose:

✔ **Cancel. Handheld will reboot:** This option makes no changes except to reboot the handheld. This is the safest option; I usually try this first.

✔ **Delete Operating System:** Choosing this option means that you will have to restore the operating system (see Chapter 4).

✔ **Delete Document Folder contents:** This is exactly like pressing the Reset button on the back of the handheld. All documents are deleted, but the OS remains intact.

✔ **Complete format (includes operating system):** This is the most powerful reset available on the TI-Nspire. All documents are deleted and the operating system will need to be restored.

If these options do not successfully fix the problem, you need to call in the experts. Call 1-800-TI-CARES and a Texas Instruments representative will guide you through the steps to repair or replace your handheld.

Avoiding Parentheses Issues

If you are on a Graphs page and type $f1(x) = 3(x - 2)$, the function graphs exactly as you expect it to. However, if you type $f2(x) = x(x - 2)$, an error message appears and the function does not graph. Why not? Both functions have implied multiplication before the parentheses.

This implied multiplication can cause other problems. For example, if I type the standard form of a quadratic equation as follows, $f3(x) = ax^2 + bx + c$, TI-Nspire recognizes bx as one variable instead of as separate variables b and *x*. If I want TI-Nspire to treat b*x* as two distinct variables, I must include a multiplication sign in between b and *x*.

Incidentally, inserting a multiplication sign before the parentheses corrects the error and the second function graphs exactly as I expected. For that reason, I always include the multiplication sign before a set of parentheses, thus avoiding the problem.

Removing a Function Table

Adding a function table to a Graphs page is easy (press ⌨ctrl⌨T). Getting rid of it can be tough if you haven't been told how. Here are four ways to accomplish this task:

- ✔ When your cursor is on the table side of the split page (notice the dark outline around the application), press ⌨ctrl⌨K to select the entire application. Press ⌨del⌨ to delete the application.
- ✔ Use the Undo command repeatedly to restore the Graphs page to a full screen.
- ✔ Move your cursor to the graph side of the screen and press ⌨. Then press ⌨ctrl⌨T again to hide your table.
- ✔ Use the ungroup shortcut command to move the table to the next page in the document. Press ⌨ctrl⌨6 to ungroup the split page into two separate pages. (I prefer this method of removing a table from a Graphs page.)

What to Do If Your Cursor Disappears

Occasionally, you may not be able to bring up a cursor on a page. If you find yourself in this situation, you don't need to panic; the case of the missing cursor can quickly be solved.

I usually try to swipe my finger on the Touchpad. If that doesn't work, try one of these fixes:

- ✔ Advance to another page (⌨ctrl⌨ ▶ or ◀) and then return to the page to restore the missing cursor.
- ✔ Access the Page Sorter view (⌨ctrl⌨▲), and then press ⌨enter⌨ to return to the Full Page view. Like magic, your cursor has been restored.

It seems that moving away from the page and returning solves the case of the missing cursor. If only all problems were this easy to fix.

Graphing a Scatter Plot

An easy fix to this problem is available as well. Usually, when students have trouble graphing a scatter plot, it is because they have failed to remember one crucial step in the process: They failed to name their lists. Here are the steps to graph a scatter plot:

1. **Type your data onto a Lists & Spreadsheet page.**

   Don't forget to move your cursor to the column/list name area located at the top of the column and type the name of each list.

2. **Insert a Graphs page, press [menu]⇨Graph Type⇨Scatter Plot, and use the entry line and the [var] button to call up the data by referencing the corresponding column name.**

3. **Press [menu]⇨Window/Zoom⇨Zoom Data to change your window so that you can have a nice view of all your data.**

   Alternatively (to Steps 2 and 3), insert a Data & Statistics page and choose the corresponding column names as the x- and y-variables.

Locating Tough-to-Find Zeros of Polynomial Functions

Sometimes, my students have trouble using the Graph Trace tool to find the zeros of a function. They try to use Graph Trace in the area of the zero, and TI-Nspire fails to identify the location of the zero. The first thing that I check is whether the operating system on the handheld is the most recent (see Chapter 4). If upgrading the OS doesn't fix the problem, a foolproof way of finding the zeros is available. While the Graph Trace tool is open, I have them press [enter], which places a point on the function with the coordinates clearly labeled. Grab and drag the point to the location of the zero, and the tough-to-find zero has been identified!

One other issue associated with finding zeros is also worth mentioning here. I graphed $f1(x) = 100x^2 - 4x - 1000$ on a Graphs page. Using the Graph Trace tool, I found a positive zero at (3.14,–8E-11). Why is the y-value (which happens to be written in scientific notation) not exactly zero? The Graphs application is an approximate environment. TI-Nspire calculated the y-coordinate of this point to be –0.00000000008, which is very close to zero. If this happens, consider the y-value to be zero.

Storing a Variable to the Wrong Name

Consider that you named a column in the Lists & Spreadsheet application, only to realize that you want to use this name as a variable elsewhere in the same problem. You try moving to the column heading and deleting the name, but the name is still a defined variable (you press var and see it still sitting there in the list of defined variables). The Undo feature doesn't work either.

To permanently delete a variable from a problem, open a Calculator page and execute the command delvar *(variable name)*. Press menu⇨Actions⇨ Delete Variable to access the delvar command.

Crowding Multiple Applications on One Page

Consider that you have three applications on one screen, and it's simply too crowded. To move one application to a separate page and reconfigure the existing page with one less application, follow these steps:

1. **Press ctrl tab until the application that you want to relocate is active.**

   Active applications are surrounded by a dark border.

2. **Press ctrl K to select the application.**

   The selected application flashes.

3. **Press ctrl X to cut the application.**

   The cut application is stored to the Clipboard.

4. **Press ctrl I to insert a new page, and press esc to remove the chosen application menu displayed in the upper-left corner of the screen.**

5. **Press ctrl V to paste the copied application to this new page.**

The pasted application appears as a full page.

Drawing the Inverse of a Function

TI-Nspire does not have a Draw Inverse command such as those on the TI-84 and TI-89. I like to use parametric graphing to accomplish this task. Insert a

Graphs page and type a function; I typed *f1(x)* = *2x* + **1**. Press menu⇨Graph Type⇨Parametric. I typed the following: *x1(t)* = *f1(t)*, *y1(t)* = *t*, and **–10** ≤ *t* ≤ **10**. One thing that I like about this method is that if you change the *f1(x)* equation, the inverse automatically updates!

If you have a TI-Nspire CAS, you could insert a Calculator page and use the Solve command to find the inverse of the equation. I typed the following: **solve(***y* = **2***x* + **1**,*x***)**. If you switch *x* and *y* in the resulting equation, you have just found your inverse.

Giving Results in Decimal Form Rather Than in Fraction Form (Or in Simplified Radical Form)

This is probably the most common problem that students experience in my classroom. It's great to have the option to see the exact form of a rational and irrational number. However, sometimes you prefer to view results as decimal approximations. In the Calculator application, press ctrl enter to force approximate results. Alternatively, include a decimal point somewhere in your expression to trick TI-Nspire into displaying a decimal approximation.

Appendix A

Safeguarding in Press-to-Test Mode

*T*I-Nspire is a powerful tool that helps students make connections in science and mathematics. When students are taking assessments, however, the teacher might want to limit the functionality of the handheld at times, to assess what students know and understand without certain TI-Nspire capabilities. Some exam agencies also allow TI-Nspire on their exams only with the condition that the Press-to-Test mode is activated.

Of course, teachers and exam agencies also have the option of making only part of their assessment calculator-active. But if you are a teacher, you may read this appendix and decide that Press-to-Test mode is a valuable tool that you could use in your own classroom. I like the fact that this tool is completely customizable. If you don't want students to be able to graph an inequality on your test, no problem. Just select that check box when you put their handhelds in Test mode.

The effects of Press-to-Test mode are only temporary. Files that were on your handheld are not deleted, but all the files in My Documents (and any Scratchpad data) are unavailable while in Test mode. Similarly, any work that was done while in Test mode will be unavailable after you exit Test mode. Can you see why Press-to-Test mode could be a valuable tool? Say it with me: "No cheating, please!"

Enabling Press-to-Test Mode

In case you forget the steps for how to activate Press-to-Test mode on the handheld, a Help icon is available that not only shows you the steps but also explains each restriction. To access the Help icon, press [doc▾]⇨Press-to-Test Mode⇨Help. Scroll through the pop-up window to look up the info.

Here are the steps to enable Test mode:

1. **Make sure that your handheld is turned off.**

2. **Press and hold both keys, [esc] [🏠on], until the Press-to-Test screen appears.**

   See Figure A-1. If you are using TI-Nspire Clickpad, press [esc], Home, and On at the same time.

3. **To accept the default settings, press [enter].**

The restrictions can be customized to fit your needs by selecting the features that you would like to activate. After you press [enter], there is no turning back — you are in Press-to-Test mode, as indicated by the Lock icon in the upper-right corner of the screen.

Figure A-1:
The Help screen and the Press-to-Test screen.

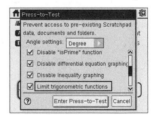

Defining the restrictions

By default, 3D graphing is disabled and all pre-existing documents, folders, and Scratchpad data are disabled. Additionally, here are the nine other restrictions that are available (see Figure A-2):

✓ **Limit geometry functions:** Press [menu] on a Geometry page and you will notice that a few items are grayed out. Selecting this check box makes the Measurement, Construction, and Transformations menus inaccessible. In addition, the powerful Coordinates & Equations command is not available in this setting.

- ✓ **Disable function grab and move:** Selecting this check box disables the ability to grab and move functions in a Graphs or Geometry application.

- ✓ **Disable vector functions**

- ✓ **Disable "isPrime" function**

- ✓ **Disable differential equation graphing**

- ✓ **Disable inequality graphing**

- ✓ **Disable trigonometric functions**

- ✓ **Disable Log template and summation functions**

- ✓ **Disable Polynomial Root Finder and simultaneous Equation Solver**

Figure A-2:
Disabled functions in Press-to-Test mode.

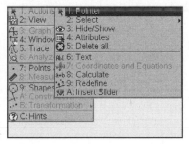 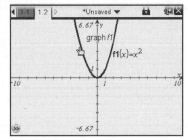

Turning the handheld on and off

If you turn your handheld off while it is in Press-to-Test mode, you can be assured that when you turn it back on, it will still be in Test mode. In fact, you will get a nice reminder screen and a quick rundown of the features that have been disabled.

Understanding the LED indicators

If you are in Press-to-Test mode, the LED indicator lights at the top of your handheld will blink, even if your handheld is turned off! The blinking lights have a purpose (besides reminding you that you are in Test mode):

- ✓ **Red-green LED indicator:** Blinks red-green-green when you are entering Test mode.

- ✓ **Yellow LED indicator:** This blinking light indicates that you are in Press-to-Test mode, but not all of the restrictions were selected.

✔ **Green LED indicator:** This blinking light indicates that you are in Press-to-Test mode, and all of the restrictions were selected.

✔ **Yellow-green-red LED indicators:** This fast blinking lasts for a little over 30 seconds as the unit reboots (getting back to its normal self). When all lights are off, you know that you have successfully exited Test mode.

Exiting Press-to-Test Mode

Information on exiting Test mode is probably the most important part of this appendix. It can be frustrating to have your handheld stuck in Press-to-Test mode. If you take out the batteries and put them back in, you are still going to be in Test mode. If you switch keypads, you will get the error message This keypad is not compatible with the current Press-to-Test mode. Some desperate student (or teacher) may even try pressing and holding the Reset button on the back of the handheld. (Please, do not do that!) Resetting the handheld is a pretty drastic step that erases all your TI-Nspire files. And guess what? — your handheld would still be in Test mode.

Here are the only four ways to exit Press-to-Test mode:

1. **Use a unit-to-unit connection cable to connect your handheld to another TI-Nspire Handheld (either a CAS or numeric TI-Nspire will work). Press [docv]⇨Press-to-Test⇨Exit Press-to-Test.**

2. **Using Connect-to-Class Teacher Software (or Navigator), have students log in, and then press [docv]⇨Press-to-Test⇨Exit Press-to-Test.**

3. **Install a different operating system.**

4. **Create a folder on your computer called "Press-to-Test"; in that folder, save an empty .tns file called "Exit Test Mode" (Folder and file are case-sensitive). Send this file to your handheld.**

Appendix B

Basic Programming

· ·

· ·

*Y*ou have two ways to write functions and programs using TI-Nspire: by typing statements directly on the Calculator entry line or by using the built-in Program Editor. The Program Editor has several advantages, including having its own application menu, which allows quick access to a variety of programming tools. Consequently, I focus strictly on using the Program Editor to write functions and programs.

In this appendix, I cover all the programming basics, including what you need to know to work with existing programs and functions as well as how to write your own programs and functions.

The Difference between Programs and Functions

Functions and programs have many similarities. However, a few differences exist. Here's how you differentiate between a function and a program:

✔ **Functions must return a result, whereas programs cannot.**

✔ **Functions can be used within mathematical expressions.** For example, consider a function called func( ) and a program called pgrm( ). The expression 4func(5) is valid, whereas 4pgrm(5) is not.

✔ **Programs can be run only from the Calculator or Notes application. Functions can be used with any application.**

✓ **A function can use another function as part of its definition or as part of its argument.** However, a function cannot refer to another program or use a program as part of its definition. Looking at the previous example, func(func(5)) is valid, whereas func(pgrm(5)) is not. However, pgrm(func(5)) works!

✓ **Programs can store local or global variables. Functions can store only local variables.**

✓ **A function can define a local function. A function cannot define a global function.**

If the information contained above leaves you a bit confused, here's another way to think about a function. A function basically takes an input(s), does some work to it, and spits out an output — which sounds a lot like the mathematical definition of function.

Working with the Program Editor Menu

The Program Editor is where you work with new or existing functions and programs. Additionally, the Program Editor has its own application menu associated with it, just like the seven core TI-Nspire applications.

Here are two ways to open the Program Editor:

✓ **Use the Tools menu.** From any page, press [ctrl][I] to insert a page and then press [doc▾]⇨Insert⇨Program Editor and select one of the first three options: New, Open, or Import.

When this method is used, the Program Editor opens in a separate page.

✓ **Use the Functions & Programs menu from within the Calculator application.** Press [menu]⇨Functions & Programs⇨Program Editor from within the Calculator application and select one of the first three options: New, Open, or Import.

When this method is used, the current page is reconfigured with one more application, and the Program Editor application is added to the current page. If four applications are already on the page, the Program Editor opens in a separate page.

I prefer to work with the Program Editor on a split page. Therefore, I almost always choose the second option. This way, I can "test" the program I write without switching pages. To accomplish this, insert a Calculator page by pressing [ctrl][doc▾]⇨Add Calculator. The first screen in Figure B-1 shows the dialog box that results from pressing [menu]⇨Programs & Functions⇨Program Editor⇨New.

I named the program volume, as shown in the second screen in Figure B-1. Program names may neither start with a digit nor contain spaces. In the dialog box, I chose the default settings of Type: Program and Library Access: None. This screen also shows the two main areas of the Program Editor view:

✔ **Status line:** This area shows the name of the current program or function and the line number that corresponds to the current cursor location. An asterisk (*) indicates that something has changed since the last time the syntax has been checked and the function or program has been stored.

✔ **Work area:** The location of the function definition or program.

Figure B-1:
The
Program
Editor
application
menu.

The third screen in Figure B-1 shows the application menu (press menu) associated with the Program Editor. Here's a brief description of the contents associated with each top-level item contained on the Program Editor application menu:

✔ **Actions:** This submenu allows you to create new functions or programs. It's also where you open, import, and view existing functions and programs as well as create copies, rename, and change the library access of existing functions and programs. Editing tools in this submenu enable you to insert comments, find/replace text, and go to a specific line.

✔ **Check Syntax & Store:** Here you check the syntax for errors and store the program or function. TI-Nspire tries to put the cursor near the first error it finds.

✔ **Define Variables:** This submenu is where you define local variables. It's also where you can insert the Func...EndFunc and Prgm...EndPgrm templates. Keep in mind that these templates are automatically inserted if you press docv ⇨Insert⇨Program Editor⇨New.

✔ **Control:** This submenu contains several functions that allow you to control the flow of a function and program. It includes conditional statement templates and looping commands.

- ✔ **Transfers:** Here are commands that allow you to terminate a function or program, jump to a different location, or alter the flow of a loop.

- ✔ **I/O:** The Input/Output submenu contains the Disp (display), Request, RequestStr (request string), and Text commands.

- ✔ **Mode:** This submenu allows you to temporarily change the mode settings associated with a function or program. For example, you can configure a function or program to display numerical results in binary form.

In this appendix, I give specific examples of how to use most of the various Program Editor application menu items.

Writing Programs

Now that you have been introduced to some of the available tools, it's time to put the tools to work writing programs.

The basic input and output of a program

Most programs look for some input when performing a task. For example, consider that you want to write a program that calculates the volume of a cylinder for the inputs *radius* and *height*.

Figure B-2 shows three different configurations of this program. To have a better view of the screens, I switched the Document view to Computer mode. Here's a brief description of how these configurations differ:

- ✔ **First screen:** I used a Calculator page on the left side of the screen to define the values of the *radius* and *height* variables. In the Program Editor, on the right side of the screen, I used the formula for the volume of a cylinder to calculate the variable vol_cyl. To open the Display command, I pressed menu➪I/O➪Disp. I enclosed text in quotes so that the program would output exactly what I had typed, and then I separated the two components with a comma followed by the variable vol_cyl. Variables (or any other calculation) should not be included in quotes if you want to see the numeric value that is associated with the variable.

- ✔ **Second screen:** Here, before you run a program, it must be stored. I pressed menu➪Check Syntax & Store➪Check Syntax & Store. A message at the top of the Program Editor says that volume was stored successfully. Oh, by the way, it is quicker to use the shortcut ctrl B to store a program.

✔ **Third screen:** Here, I use a Calculator page to run the program volume(). Because I used one Display command with a comma to separate the string and the variable, they both appear on the same output line.

In this first example, the *radius* and *height* were defined before the program was run.

Figure B-2:
Using input
and output
statements
in a
program.

Use the Text command to make your output stand out! Executing the command Text "Remember, logarithms are exponents!" pauses the program until you press OK to resume execution of the program. See Figure B-3.

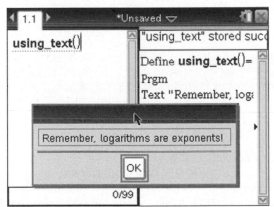

Figure B-3:
Using text.

How to input values using the argument and the Request command

You have two more-sophisticated methods to input values into a program. When the program volume is run, you can *pass the variables* to the program by entering the values of the variables *radius* and *height* in the argument.

For example, the command `volume(3,5)` calculates the volume of a cylinder with radius 3 and height 4. Of course, the program needs to be adjusted accordingly to accommodate this method of input. See the first screen in Figure B-4. I also changed the way the output displayed. I used the Display command and put the formula in quotes on the first line of output. On the second line of output, I wanted to express the exact volume, so I evaluated (*radius*²)(*height*) and put everything else in quotes to display as a string.

My favorite way to input values into a program is by using the Request command. Press [menu]⇨I/O⇨Request to open the Request tool. Using this command, a dialog box pops up, while the program is running, to request information. See the second screen in Figure B-4. Notice in the third screen, the answers are pasted into the Calculator page when you press [enter].

Figure B-4:
Inputting values using the argument and the Request command.

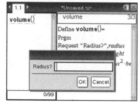

Don't forget to press [ctrl] [B] to invoke the Check Syntax & Store command each time you make changes to your program. An * in front of the name of the program in the Program Editor indicates that changes have been made to the program.

Using local variables

The variable `vol_cyl` can be accessed from within any application within the same problem, and its current value is determined by the value assigned to it the last time that the program `volume` was run. To see the value of `vol_cyl` in the Calculator application, press [var], select `vol_cyl`, and press [enter].

Sometimes, it's advantageous to define a *local variable* within a program. Local variables are variables that cannot be accessed from outside a program. It's a good idea to use local variables when they are used as counters in a program. In such cases, there's no reason to make this variable accessible outside a program — its only purpose resides within the program.

To specify a local variable, press menu⇨Define Variables⇨Local from within the Program Editor. Next, type the name of the variable. For example, Local *x* defines *x* as a local variable.

Controlling the flow of a program

Typically, programs or functions execute commands in sequential order. However, sometimes you want to redirect the flow of the program. In the following sections, I talk about how conditional statements (the Lbl [label] and Goto [go to label]) commands and loops can alter the flow of a program and enhance the efficiency and functionality of a program or function.

Using conditional statements

The Control submenu of the Program Editor application menu includes conditional commands. These commands test a condition and, based on the result, decide which part of a program to execute.

If you only want to execute a single command based on a true condition, press menu⇨Control⇨If to access the If command. The following is a simple example that uses the If command:

```
If x > 10
Disp "x is greater than 10"
x:=x+1
```

In this example, the program displays x is greater than 10 if the current value of *x* is greater than 10. If not, line 2 is skipped and the program moves on and executes the command in line 3.

Keep in mind that you must store a value to *x* before executing the If command.

If you only want to execute multiple commands based on a true condition, press menu⇨Control⇨If...Then...EndIf to insert an If...Then...EndIf structure. The following is a simple example that uses this command:

```
If remain(x,3)=0 Then
  Disp "x is divisible by 3"
  Disp "The sum of the digits are also divisible by 3"
EndIf
```

In this example, the program tests to see whether *x* divided by 3 has a remainder of 0. If this condition is met, the block of commands that precede EndIf is executed. Otherwise, the program skips to the line immediately after EndIf.

Sometimes, you want to execute one command for a true result and a different command for a false result. In such cases, press [menu]⇨Control⇨If...Then...El...EndIf to insert an `If...Then...Else...EndIf` structure. Take a look at the example that follows:

```
If x>10 Then
   Disp "x is greater than 10"
Else
   Disp "x is less than or equal to 10"
EndIf
Disp x
```

Consider that *x* initially has a value greater than 10. The program executes lines 2, skips lines 3 through 5, and executes line 6. If *x* initially has a value less than or equal to 10, the program skips line 2, executes line 4, and executes line 6.

To perform multiple tests, consider using an `If...Then...ElseIf...EndIf` structure. Here's an example in which I test for divisibility by three different numbers:

```
If remain(x,2)=0 Then
   Disp "x is divisible by two"
ElseIf remain(x,5)=0 Then
   Disp "x is divisible by five"
ElseIf remain(x,7)=0 Then
   Disp "x is divisible by seven"
EndIf
```

Press [menu]⇨Control⇨ElseIf...Then to insert an `ElseIf...Then` structure.

Using the Lbl and Goto commands to jump to different locations in a program

The Lbl and Goto commands work in conjunction to direct the program to jump from one section to another. You can type these commands using the keypad or access them from the Transfers submenu (press [menu]⇨Transfers⇨Lbl and press [menu]⇨Transfers⇨Go to Lbl).

Because the Goto command is not conditional, it is often used in conjunction with an If command. This way, the Goto command is only executed if a certain condition is met. Check out the simple example that follows:

```
If x < 10 Then
Goto line5
Disp "The number is greater than or equal to 10."
Stop
Lbl line5
Disp "The number is less than 10"
```

In this example, if $x < 10$, the program executes line 2, jumps to line 5, and continues on from there. If $x \geq 10$, the program skips line 2 and executes lines 3 and 4 (line 4 stops the program).

TIP

You can use multiple Lbl and Goto commands within the same program. Just make sure that pairs of Lbl and Goto commands have the same label. Follow standard variable-naming conventions with the Lbl and Goto commands.

Using loops

To repeat a set of commands in succession, use one of the loop commands found on the Control submenu (press menu⇨Control). Each of the available loop commands uses a conditional test (often with a counter) to determine when to exit a loop.

The For...EndFor loop uses a counter to control the number of times that the loop is repeated. Press menu⇨Control⇨For...EndFor to paste the For... EndFor structure to the cursor location. Here's an example of how this loop is used:

```
For x,0,10,2
    Disp 3·x+1
EndFor
Disp x
```

The syntax associated with the first line is For variable, begin, end, [increment]. I set up this loop to start at $x = 0$, increment by 2, and exit the loop when x exceeds 10. If I omit the increment value, the increment is assumed to be 1. This loop displays the values 1, 7, 13, 19, 25, and 31, followed by 12 — the value of x that finally breaks the loop.

The While...EndWhile loop repeats a block of commands as long as a condition remains true. Press menu⇨Control⇨While...EndWhile to paste the While...EndWhile structure to the cursor location. Here's an example of how this loop is used:

```
x:=1
While x<5
  Disp x+3
  x:=x+1
EndWhile
Disp x
```

Because x is initially 1, the two commands that precede EndWhile are executed. The second command increases x by 1. When x reaches a value of 5, the loop breaks and the program executes line 6. The program displays 4, 5, 6, 7, 5. The first four values come from the Display command contained in the loop. The last value comes from the Display command located in line 6.

The `Loop...EndLoop` command sets up an infinite loop. The only way to break this type of loop is to include a command such as If, Exit, Stop, or Goto. Press menu⇨Control⇨Loop...EndLoop to paste the `Loop...EndLoop` structure to the cursor location. Here's an example of how this loop is used:

```
x:=1
Loop
  If x<5 Then
    Disp 2·x
    x:=x+1
  Else
    Exit
  EndIf
EndLoop
Disp x
```

Here, I initially store 1 to the variable x. After I enter the loop, the program checks to see whether x is less than 5. If so, it displays the value of $2x$ and increments x by 1. This continues until $x = 5$, at which point the program executes the Exit command and breaks the loop. The Exit command takes the program to the line immediately following `EndLoop`. The program displays 2, 4, 6, 8, 5. The first four values come from the Display command contained in the loop. The last value comes from the Display command located after `EndLoop`.

The Exit command can be used to break any loop that uses a `For...EndFor`, `While...EndWhile`, or `Loop...EndLoop` structure.

Putting It All Together: A Sample Program

In the previous sections of this chapter, I give you little snippets of code for the purpose of illustrating how to use a particular command or sequence of commands when writing programs. The real trick when writing programs is to combine a series of commands in a logical format for the purpose of performing more complex and sophisticated tasks. The example given in this section certainly is not the end-all. Rather, it is intended to show how some of the commands covered in this chapter can be used to perform a task that may be of interest to you.

Listing B-1 shows the code for a program called `guess_a_number`. This program picks a random number between 0 and 100. Your goal is to guess the mystery number in as few attempts as possible.

Listing B-1:   The guess_a_number program

```
Define guess_a_number()=
Prgm
n:=randint(1,99)                                        3
Local Attempts
Attempts:=0
Text "Guess an integer between 0 and 100"
Lbl begin                                               7
Request "Guess?",number                                 8
attempts:=attempts+1
If number>n Then                                        10
 Disp "too high, try again
 Goto begin
ElseIf number<n Then                                    13
 Disp "too low, try again"
 Goto begin
Else                                                    16
 Disp "You guessed it!"
 Disp "It took you",attempts,"tries"
EndIf
EndPrgm
```

Here's a description of each of the numbered regions of the program:

3 Here, I use the Random Integer command to randomly select an integer from 1 to 99 inclusive. I then define the variable Attempts as a local variable (recall that this is a good idea when using a variable to "count"). I define Attempts at 0, because no attempts have been made. Using the Text command, I pause the program to give instructions to the user.

7 This section is marked by a Label command, which I can use to loop through each time an answer is incorrect.

8 Here, I ask the user to type in a guess. Each time, I add one to the "count" by redefining attempts. Notice that this statement is inside the loop.

10 Here is the conditional If statement that initially confirms whether the guessed number is larger than the mystery number, *n*. If true, *both* the Display and Goto commands are executed.

13 A continuation of the conditional If statement, using Elseif determines whether the guessed number is less than the mystery number, *n*. Again, both statements that follow are executed if the Elseif statement is true.

16 Because only one other option is possible, Else is a good choice here. The user's guess and the mystery number match! Notice that attempts (a calculated value) is not in quotes when using the Display command.

The first screen in Figure B-5 shows a result when `guess_a_number()` is executed. The second screen shows the Request dialog box asking for another guess. The third screen shows the result of correctly guessing the mystery number.

Figure B-5:
The
`guess_a_`
`number`
program
results.

 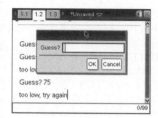 

Changing the mode settings from within a program

The `setMode( )` function uses two numerical inputs to temporarily set specific mode functions. Here's how to use this function:

1. **Position the cursor at a point where you want to insert the** `setMode( )` **command.**

2. **Press** [menu]⇨**Mode to view a list of mode settings.**

   This list includes Display Digits, Angle, Exponential Format, Real or Complex, and so on.

3. **Highlight a mode setting and press** [enter] **to view the specific settings.**

4. **Select a setting and press** [enter] **to paste the command to the cursor location.**

   For example, press [menu]⇨Mode⇨Auto or Approx⇨Approximate to set the program or function to display results in decimal form. The syntax associated with this action is `setMode(5,2)`.

Keep in mind that mode changes made within a program or function do not affect the document or system settings.

Working with Existing Programs or Functions

You can access a public or private library function or program from any document. Any function or program that is not defined as a library object can only be accessed from within the same problem in which it was created.

Contained within this section are all the options available with existing programs and functions:

✔ **Opening an existing program or function:** To open a program or function that has previously been defined in the current problem, press [doc▾]⇨ Insert⇨Program Editor⇨Open. Or, from the Calculator application menu, press [menu]⇨Functions & Programs⇨Program Editor⇨Open.

Select a function or program from the list of available items. Use this option if you want to edit a program or function that resides in the current problem.

✔ **Importing an existing program or function:** To import a program or function that is defined as a library object, press [doc▾]⇨Insert⇨Program Editor⇨Import. Or, from the Calculator application menu, press [menu]⇨Functions & Programs⇨Program Editor⇨Import.

A dialog box opens. Select the name of the library document (see the first screen in Figure B-6) that was used to define the library object. Then select the library object name, as shown in the second screen in Figure B-6. I am importing the library object `factors` from the library document `add`. After pressing [enter], I get the image shown in the third screen in Figure B-6.

Figure B-6: Importing an existing library object.

 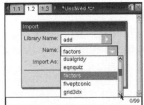 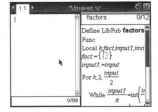

See Appendix C for more information about library objects.

- **Viewing an existing program or function:** To view a program or function that has been previously defined in the current problem or as a library object, press [doc▾]⇨Insert⇨Program Editor⇨View. Or, from the Calculator application menu, press [menu]⇨Functions & Programs⇨Program Editor⇨View.

  This option does not allow editing of a program or function. Rather, the program or function opens in a separate dialog box for viewing only.

- **Making a copy of an existing program or function:** Sometimes, it's easier to modify an existing program or function rather than create one from scratch. Follow these steps to perform this task:

  a. Open or import an existing program or function in the Program Editor.

  b. Press [menu]⇨Actions⇨Create Copy.

  c. Type a new name in the Save As field and press [enter] to create the copy.

  d. Edit the program or function as needed.

  The new copy is saved to the current document. If the copy came from a library object and you want the edited copy to remain a library object, you must save the current document to the MyLib folder.

- **Renaming an existing program or function:** To rename an existing program or function, follow these steps:

  a. Open or import an existing program or function in the Program Editor.

  b. Press [menu]⇨Actions⇨Rename.

  c. Type a new name in the Rename As field.

  d. If desired, change the access level by selecting the appropriate setting from the Library Access drop-down menu.

  e. Press [enter] to close the dialog box and put the changes into effect.

- **Changing the library access settings of an existing program or function:** To change the library access settings of an existing program or function, follow these steps:

  a. Open or import an existing program or function.

  b. Press [menu]⇨Actions⇨Change Library Access.

  c. Select the appropriate setting from the Library Access drop-down menu.

  d. Press [enter] to close the dialog box and put the changes into effect.

If you change the library access settings to LibPriv or LibPub (Show in Catalog), you must save the current document to the MyLib folder for your function or program to be available as a library object.

✔ **Closing an existing program or function:** To close an open program or function, follow these steps:

    a. Press menu⇨Actions⇨Close.

    b. If you have not already done so, you are greeted by the prompt Do you wish to check the syntax and store? Press enter to check the syntax and store or press tab enter to close the program or function without checking the syntax and storing.

If you close a function or program without storing, all your work is lost.

✔ **Running an existing program or function:** You can run a function from any application. You can only run a program from the Calculator or Notes application.

To run a program or function defined within the current problem, follow these steps:

    a. Press var to access a list of current problem functions and programs.

    b. Select a program or function, and press enter to paste it into the current application.

       You can also type the program or function name using the keypad.

    c. Include parentheses after the name and include one or more arguments, if needed, within the parentheses.

       You must include parentheses after the function or program name, even if it does not require an argument.

    d. Press enter to run the program or function.

See Appendix C for information on how to run a function or program defined as a library object.

A New Way to Execute a Program

I have mentioned this a couple of times, but maybe you missed it. You can only run a program from the Calculator or Notes application. Wait, you can run a program in the Notes application? Yes, you can! And, you have a particularly nice advantage to doing so. If you're in a math box on a Notes page and run a program, it will continuously update. This opens the floodgates of possibilities of what can be accomplished with a program on TI-Nspire!

Let me show you one use that advanced authors have employed to make their documents better. In the first screen in Figure B-7, I have inserted a Graphs page, turned the grid on, placed a point on the grid, and stored the *x*- and *y*-coordinates as *xc* and *yc*.

In the second screen in Figure B-7, I have written a program named `con-tain`. I have used piecewise functions (that work like an if-then statement in reverse) to contain the possible values of *xc* and *yc*. The piecewise function for *xc* says that if *x<–9* then *x:=–9*, if *x>9* then *x:=9*, else the *xc* value is stored unchanged.

The third screen executes the program in a math box. Notice that the result, Done, confirms that the program is running, continuously.

What does this program do? The point on the grid cannot be moved off the screen. If you grab and move the point, the program assures that –9<*xc*<9 and –6<*yc*<6.

Figure B-7:
Executing a program on a Notes page.

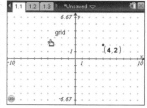 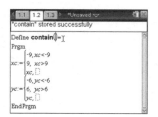 

A Note about Lua Scripting

Texas Instruments opened up the possibilities when they upgraded TI-Nspire to accept Lua programming. Lua scripting is much more powerful than simply accessing the programming capabilities of TI-Nspire. Lua takes some time to absorb, so if you have a programming bent and want to make powerful interactive documents, you will need to do some exploration on your own. There is no better place to start than Stephen Arnold's website: `http://compasstech.com.au/TNS_Authoring/Scripting/`.

Appendix C

Working with Libraries

*I*n this appendix, I talk about how libraries provide global access to a function, variable(s), or program. I also talk about the difference between public and private libraries and describe how a library is created. Finally, I show you the steps used to call a library into play from any open document.

How Libraries Improve Efficiency

Libraries are TI-Nspire documents that reside in the MyLib folder of My Documents. Libraries contain variables, functions, or programs that are available from within any open document. These are different from ordinary variables, functions, and programs, which are available only from within a single problem within a single document.

If you find yourself defining the same function, storing the same variables, or performing a series of steps repeatedly, you may want to consider defining a library object. For example, you can create a program defined as a library object that takes the inputs a, b, and c and uses the discriminant to give information about the number and nature of the solutions to an equation of the form $0 = ax^2 + bx + c$. By defining this program as a library object, it is available in any document, not just the document in which it was created. Using the discriminant is a fairly common task and one that warrants defining the program as a library object.

Here's an overview of the process used to create private and public library objects:

1. **Open a new document and define one or more private or public library objects.**

   Select a name for each library object that helps describe its purpose.

2. **Save the document to the** `MyLib` **folder.**

   Use a name that describes the library objects contained within the document. This document is referred to as a *library document.*

3. **Open a new or existing document.**

4. **From the Calculator application, press** menu⇨**Actions**⇨**Library**⇨**Refresh Libraries to refresh the library list.**

5. **Type the long name of the private library object or use the Catalog to quickly access a public library object.**

   The Calculator or Notes application can run library objects that are programs. You can access variables or functions defined as library objects from any application.

Creating Public and Private Libraries

A *public library object* is a variable, function, or program that appears in the Catalog. A *private library object* is one that does not appear in the Catalog and is typically used to perform basic, low-level tasks. If you don't want anyone else to use your program, define your program as a private library object.

Consider that you want to define a public library object that enables you to find the area of any triangle using Heron's formula. Heron's formula states that the area of a triangle, *A*, with sides *a*, *b*, and *c*, is given by the formula

$$A = \sqrt{s(s-a)(s-b)(s-c)}$$

where *s* is the semi-perimeter defined as

$$A = \frac{a+b+c}{2}$$

In this section, I create a public library program called `heron` that calculates the area of a triangle. Within this program, I define a private library function `sp` that calculates the semi-perimeter *s*.

Here are the steps used to create the private library function sp:

1. **Open a new document with a Calculator page.**

2. **Press** menu⇨**Actions**⇨**Library**⇨**Define LibPriv.**

3. **Type the command shown in the first screen in Figure C-1.**

   This creates a function that takes the lengths of the sides of a triangle, *a*, *b*, and *c*, and determines the semi-perimeter.

At this point, I have the option of saving the current document to the MyLib folder, the location to which all library objects must be saved. However, I have chosen to first define the public library program heron and then save the document.

When defining a public library object that is a variable or a function, press menu⇨Actions⇨Library⇨Define LibPub (Show in Catalog) from within the Calculator application and then define the variable or function. To create a public library object that is a program (which is true in this situation), follow these steps:

1. **Press** docv⇨**Insert**⇨**Program Editor**⇨**New to open the Program Editor dialog box.**

   You can also press menu⇨Functions & Programs⇨Program Editor⇨New.

2. **Type the name of the program.**

   I use heron for the name of the program.

3. **Configure the Library Access field for LibPub (Show in Catalog).**

   I also have the option to save my program as a private library object.

4. **Press** enter **to close the dialog box and open the Program Editor in a new Calculator page.**

   If you press menu⇨Functions & Programs⇨Program Editor⇨New to access the Program Editor, the Program Editor opens in a split page.

5. **Type the program as shown in the second screen in Figure C-1.**

   The third line from the bottom of the heron program, in its entirety, is

$$A = \sqrt{sp(a, b, c)\big(sp(a, b, c) - a\big)\big(sp(a, b, c) - b\big)\big(sp(a, b, c) - c\big)}$$

   Notice the private library function sp(a,b,c) is called into play for the purpose of calculating the semi-perimeter. I certainly could have calculated the semi-perimeter within this program but wanted to show you how a library object can be accessed from within another library object.

A LibPub program can display a syntax line at the bottom of the Catalog when it is highlighted. All you have to do is insert a comment (press menu⇨Actions⇨Insert Comment) in the first line of the program.

See Appendix B for more information about programming.

6. **Press** menu⇨**Check Syntax & Store**⇨**Check Syntax & Store (or use the shortcut key sequence** ctrl B**).**

   This step is important. Saving the document does not automatically store the program.

7. **Save the document to the** MyLib **folder.**

   As you can see in the third screen in Figure C-1, I have named my document geo_formula. This document contains both library objects. I have chosen to define these library objects in the same library document because they are related to one another.

Figure C-1: Creating private and public library objects.

 Use the GetVarInfo command from within the Calculator application of a library document to view a list of all stored variables, including those that have been defined as library objects. This is particularly helpful if you did not create the library document and want to view the library objects contained within it.

 Library documents can also be created using TI-Nspire Computer Software. The default save locations for computer-based library documents are My Documents\TI-Nspire\MyLib (Windows) and Documents\TI-Nspire\MyLib (Macintosh).

Using Library Objects

In the following sections, I show you how to access privately and publicly defined library objects from any open document.

Private libraries

To access the private library object sp(a, b, c), follow these steps:

1. **Open a new or existing document.**

2. **Press [doc▾]⇨Refresh Libraries.**

   This refreshes the libraries and provides access to those libraries that have been previously defined.

3. **Type the name of the document that contains the private library object, the backslash symbol (\), the name of the private library object, and the argument(s) for the private library object, if necessary.**

   Referring to the first screen in Figure C-2, geo_formula is the name of the document that was used to create the private library object sp(a, b, c). Recall that this library object requires three inputs that are the sides of the triangle.

   Press [menu]⇨Actions⇨Library⇨Insert "\" Character to insert the backslash symbol.

4. **Press [enter] to execute the command.**

The first screen in Figure C-2 shows that the semi-perimeter of a triangle with sides 5, 6, and 7 is 9.

Public libraries

You can access the public library object heron(a, b, c) from any open document. First, if you haven't already done so, choose [doc▾]⇨Refresh Libraries to refresh the libraries. You then have two options:

✔ **Use the long name. You can do this as follows:**

   a. Open a Calculator page and type **geo_formula\heron(a,b,c)**, where *a*, *b*, and *c* are the sides of the triangle.

   b. Press [enter] to execute the command.

✔ **Use the Catalog. You can do this as follows:**

   a. Open a Calculator page and press [⌂][5] to access the public libraries (press [⌂][6] if you are using TI-Nspire CAS).

   b. Highlight geo_formula and press [?] to reveal its contents.

      All public libraries defined in the document geo_formula are listed here.

    c. Highlight `heron` and press `enter` to paste it into the Calculator application.

       See the second screen in Figure C-2. Notice that the comment I inserted in the program shows up in the Catalog as syntax "help."

    d. Type the argument(s) for the program, if necessary, and press `enter` to execute the command.

The third screen in Figure C-2 shows that a triangle with sides 5, 6, and 7 has an area of approximately 14.6969.

Figure C-2:
Using
private
and public
library
objects.

If the current open document is the same document that was used to create a library object, you can access this library object via its short name. For the example used in this chapter, the short name for the private library object is `sp(a,b,c)` and the short name for the public library object is `heron(a,b,c)`. The long names for these library objects are `geo_formula\sp(a,b,c)` and `geo_formula\heron(a,b,c)`, respectively.

Index

• *G* •

• P •

Notes

Notes

Apple & Macs

iPad For Dummies
978-0-470-58027-1

iPhone For Dummies,
4th Edition
978-0-470-87870-5

MacBook For Dummies, 3rd
Edition
978-0-470-76918-8

Mac OS X Snow Leopard For
Dummies
978-0-470-43543-4

Business

Bookkeeping For Dummies
978-0-7645-9848-7

Job Interviews
For Dummies,
3rd Edition
978-0-470-17748-8

Resumes For Dummies,
5th Edition
978-0-470-08037-5

Starting an
Online Business
For Dummies,
6th Edition
978-0-470-60210-2

Stock Investing
For Dummies,
3rd Edition
978-0-470-40114-9

Successful
Time Management
For Dummies
978-0-470-29034-7

Computer Hardware

BlackBerry
For Dummies,
4th Edition
978-0-470-60700-8

Computers For Seniors
For Dummies,
2nd Edition
978-0-470-53483-0

PCs For Dummies,
Windows
7 Edition
978-0-470-46542-4

Laptops For Dummies,
4th Edition
978-0-470-57829-2

Cooking & Entertaining

Cooking Basics
For Dummies,
3rd Edition
978-0-7645-7206-7

Wine For Dummies,
4th Edition
978-0-470-04579-4

Diet & Nutrition

Dieting For Dummies,
2nd Edition
978-0-7645-4149-0

Nutrition For Dummies,
4th Edition
978-0-471-79868-2

Weight Training
For Dummies,
3rd Edition
978-0-471-76845-6

Digital Photography

Digital SLR Cameras &
Photography For Dummies,
3rd Edition
978-0-470-46606-3

Photoshop Elements 8
For Dummies
978-0-470-52967-6

Gardening

Gardening Basics
For Dummies
978-0-470-03749-2

Organic Gardening
For Dummies,
2nd Edition
978-0-470-43067-5

Green/Sustainable

Raising Chickens
For Dummies
978-0-470-46544-8

Green Cleaning
For Dummies
978-0-470-39106-8

Health

Diabetes For Dummies,
3rd Edition
978-0-470-27086-8

Food Allergies
For Dummies
978-0-470-09584-3

Living Gluten-Free
For Dummies,
2nd Edition
978-0-470-58589-4

Hobbies/General

Chess For Dummies,
2nd Edition
978-0-7645-8404-6

Drawing
Cartoons & Comics
For Dummies
978-0-470-42683-8

Knitting For Dummies,
2nd Edition
978-0-470-28747-7

Organizing
For Dummies
978-0-7645-5300-4

Su Doku For Dummies
978-0-470-01892-7

Home Improvement

Home Maintenance
For Dummies,
2nd Edition
978-0-470-43063-7

Home Theater
For Dummies,
3rd Edition
978-0-470-41189-6

Living the
Country Lifestyle
All-in-One
For Dummies
978-0-470-43061-3

Solar Power Your Home
For Dummies,
2nd Edition
978-0-470-59678-4

Available wherever books are sold. For more information or to order direct: U.S. customers visit www.dummies.com or call 1-877-762-2974.
U.K. customers visit www.wileyeurope.com or call (0) 1243 843291. Canadian customers visit www.wiley.ca or call 1-800-567-4797.

Internet

Blogging For Dummies,
3rd Edition
978-0-470-61996-4

eBay For Dummies,
6th Edition
978-0-470-49741-8

Facebook For Dummies,
3rd Edition
978-0-470-87804-0

Web Marketing
For Dummies,
2nd Edition
978-0-470-37181-7

WordPress
For Dummies,
3rd Edition
978-0-470-59274-8

Language & Foreign Language

French For Dummies
978-0-7645-5193-2

Italian Phrases
For Dummies
978-0-7645-7203-6

Spanish For Dummies,
2nd Edition
978-0-470-87855-2

Spanish
For Dummies,
Audio Set
978-0-470-09585-0

Math & Science

Algebra I
For Dummies,
2nd Edition
978-0-470-55964-2

Biology For Dummies,
2nd Edition
978-0-470-59875-7

Calculus For Dummies
978-0-7645-2498-1

Chemistry For Dummies
978-0-7645-5430-8

Microsoft Office

Excel 2010 For Dummies
978-0-470-48953-6

Office 2010 All-in-One
For Dummies
978-0-470-49748-7

Office 2010 For Dummies,
Book + DVD Bundle
978-0-470-62698-6

Word 2010 For Dummies
978-0-470-48772-3

Music

Guitar For Dummies,
2nd Edition
978-0-7645-9904-0

iPod & iTunes For
Dummies, 8th Edition
978-0-470-87871-2

Piano Exercises
For Dummies
978-0-470-38765-8

Parenting & Education

Parenting For Dummies,
2nd Edition
978-0-7645-5418-6

Type 1 Diabetes
For Dummies
978-0-470-17811-9

Pets

Cats For Dummies,
2nd Edition
978-0-7645-5275-5

Dog Training For Dummies,
3rd Edition
978-0-470-60029-0

Puppies For Dummies,
2nd Edition
978-0-470-03717-1

Religion & Inspiration

The Bible For Dummies
978-0-7645-5296-0

Catholicism For Dummies
978-0-7645-5391-2

Women in the Bible
For Dummies
978-0-7645-8475-6

Self-Help & Relationship

Anger Management
For Dummies
978-0-470-03715-7

Overcoming Anxiety
For Dummies,
2nd Edition
978-0-470-57441-6

Sports

Baseball
For Dummies,
3rd Edition
978-0-7645-7537-2

Basketball
For Dummies,
2nd Edition
978-0-7645-5248-9

Golf For Dummies,
3rd Edition
978-0-471-76871-5

Web Development

Web Design
All-in-One
For Dummies
978-0-470-41796-6

Web Sites
Do-It-Yourself
For Dummies,
2nd Edition
978-0-470-56520-9

Windows 7

Windows 7
For Dummies
978-0-470-49743-2

Windows 7
For Dummies,
Book + DVD Bundle
978-0-470-52398-8

Windows 7 All-in-One
For Dummies
978-0-470-48763-1

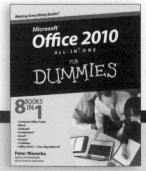

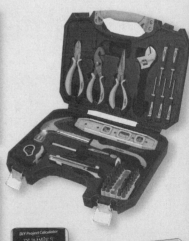

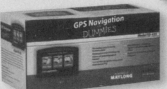